MODERN
MYTHOLOGY
AND
SCIENCE

MODERN MYTHOLOGY

AND

SCIENCE

DAN BRASOVEANU, PH. D.

iUniverse, Inc.

New York Bloomington Shanghai

Modern Mythology and Science

iUniverse books may be ordered through booksellers or by contacting:

iUniverse
1663 Liberty Drive
Bloomington, IN 47403
www.iuniverse.com
1-800-Authors (1-800-288-4677)

Because of the dynamic nature of the Internet, any Web addresses or links contained in this book may have changed since publication and may no longer be valid.

ISBN: 978-0-595-48499-7 (pbk)
ISBN: 978-0-595-60591-0 (ebk)

Printed in the United States of America

It is the perfection of God's works that they are all done with the greatest simplicity. He is the God of order and not of confusion.

—Sir Isaac Newton

Contents

1

Crisis in Modern Physics

Until the beginning of the 20th century, all branches of physics were based on the classic paradigm established by Sir Isaac Newton. According to this paradigm, space is void, three dimensional and Euclidian. There is a preferred frame of reference. Velocity measured with respect to this frame is absolute. Time is separate from space and universal. Objects have a well defined position and momentum and obey strictly deterministic laws. Waves and particles are separate entities. The Classic Law of velocity addition is always applicable; therefore the speed of light in vacuum is not the same with respect to all inertial frames of reference.

The Michelson and Morley (MM) experiment led to a new paradigm for space and time. According to Einstein's Special Relativity Theory (SRT), there is no preferred frame of reference; no absolute velocity and universal time must be replaced by local times, which depend on reference frames. The speed of light in vacuum is the same in any inertial frame. The General Relativity Theory (GRT) goes even further. Time is now treated as the fourth dimension of space. Space is considered curved not Euclidian. But physicists did not stop at only four dimensions. According to string theory, space has more then 10 dimensions. Most dimensions are twisted like pretzels and cannot be detected. Furthermore, string theory has numerous flavors. According to a more recent one, space has 26 dimensions. But wait a minute; this information is at least one year old. Surely by now, experts are poised to push through the 30 dimensions barrier, even the 40 dimensions barrier. Therefore laymen may be perhaps excused for harboring doubts about relativity and string theory.

Advances in quantum physics led to yet another paradigm. According to the interpretation of Quantum Mechanics promoted by the Copenhagen School (led by Bohr, Heisenberg, Born, Pauli and Neumann), which is now the established foundation of quantum physics and will be called QM for short, definite position and velocity and strictly deterministic laws are flawed concepts. A state vector is the most accurate description of particle position and velocity. A state vector is a set of numbers providing the probability of detecting a particle and a certain particle velocity at a given location and time. According to QM, quantum particles obey the roll of dice not deterministic laws. Furthermore, "elementary" particles are waves at the same time, etc.

These developments caused a crisis in physics. Nature is unitary. Nevertheless, contemporary physicists rely on conflicting paradigms to describe various aspects of nature. Classical Physics offers clear, deterministic predictions; QM offers only statistical predictions. Nevertheless numerous attempts to reconcile these two paradigms are based on the belief that QM reveals the most fundamental laws of nature and Classical Physics is just an approximation with limited scope. According to this belief, macroscopic objects only seem to have a definite position and velocity and to obey deterministic laws—after all, the position and velocity of these objects also exhibit some uncertainty. Position and velocity uncertainty should be considered inherent properties of all objects, not the result of some neglected factors. This view leads to some interesting scenarios, such as Schrodinger's cat. Put a live cat, some radioactive material a Geiger detector and a canister with a deadly poison in a closed cage. The amount of radioactive material is carefully measured to make sure the probability of triggering the Geiger detector once per hour is 50%. If the detector is triggered, the poison is released from canister and the cat dies. Therefore the odds of finding a dead or a live cat are even if the cage is opened after an hour. According to common sense, the caged animal is either dead or alive. But according to the QM paradigm, this simplistic conclusion must be rejected because the laws of nature are not deterministic. The cat is both alive and dead at the same time, until the cage is opened and the cat's wave function collapses. This scenario could lead a layperson to the conclusion that QM and Classical Physics should be reconciled by looking for flaws in the former not in the latter. Even a few physicists shared this simple-minded conclusion and dared to find a deterministic explanation of quantum phenomena. The scientific establishment quickly rejected such attempts. After all, Von Neumann, Kochen, Specker, Bell et al have demonstrated that no deterministic model can explain quantum phenomena. Or did they not?

2

The Theory of Internal Energy

All previous attempts to reconcile QM and Classical Physics have failed miserably because no one has asked the fundamental question: which paradigm violates the fundamental laws of physics? The answer to this question will come as a rude shock to the scientific establishment.

Nomenclature

a_n the acceleration of the nth "elementary" particle

c average amplitude of internal oscillations

E_{int} energy of internal oscillations

E_{total} total energy

f function

H system Hamiltonian

h Planck's constant

j summation index

K.E. kinetic energy

k spring constant

k_1 l^{th} component of the propagation vector

L size of "elementary" particle

m mass of "elementary" particle

n_1 summation index

p probability of detection

$\vec{p}$ momentum of "elementary" particle

˘ period of internal oscillations

time

J the quantum potential postulated by D Bohm

V̌ potential energy due to surrounding objects

√' equivalent potential

˙ speed of "elementary" particle

ᴋ, y, z Cartesian coordinates

r̃ position vector

Greek

Δl deflection

Φ total potential energy of an "elementary" particle such as electron

Φ_{sys} total potential energy of a multi-particle system

ρ Fourier coefficient

<ᴋ> average value of κ

ξ_1, ξ_2,...ξ_n arguments

σ stress

ψ displacement function

ω angular frequency of internal oscillations

2.1. QM and the Energy of "Elementary" Particles

According to QM an "elementary" particle can have only two types of energy: potential energy due to fields generated by other particles (for example, in an isolated hydrogen atom, the potential energy of the electron is due to the electromagnetic field generated by the proton) and kinetic energy. Therefore, the total energy of an "elementary" particle is given by:

$$E_{total} = V + K.E. \qquad (2\text{-}1)$$

where V is potential energy due to surrounding objects and K.E. is the kinetic energy of particle. Equation (2-1) is the bedrock of Q.M. This equation, is

explicitly reproduced or implicitly assumed in the entire QM literature and i.
dead wrong.

2.2. Internal Energy

No composite structure is perfectly rigid. Due to deformation, any structure
accumulates internal energy, which cannot be neglected. Bridges flex down
under the weight of a heavy truck and straighten up when the vehicle gets off.
The struts of a car are compressed when the wheels hit a pothole and then
rebound due to internal energy.

What happens when internal energy and the associated material fatigue are
neglected or improperly calculated? Here are just a few examples: bridges
collapse, aircraft disintegrate in flight like the ill-fated De Havilland Comets,
foam hits the heat tiles or the leading edge of a wing and space shuttles are
destroyed.

In most cases, internal energy, which is accumulated as a result of
deformation, forces the material of an object to vibrate in a harmonic manner.
Such vibrations are internal phenomena. Grip a slinky with both hands, keep it
vertical and stretch the coils forcefully, see Fig. 2-1. Then release the lower end.
The slinky coils will bounce up and down; see Fig. 2-2 and Fig. 2-3. Consider
the total energy of the slinky. Fields created by surrounding objects do not
provide potential energy because only Earth's gravity is important and is
countered by the upper hand. The slinky did not move prior to release, therefore
had no kinetic energy and according to equation (2-1), the total energy of the
slinky is zero. Nevertheless, the slinky bounces rapidly up and down after
release. The fundamental equation of QM indicates that slinky energy appears
out of nothing. Is there perhaps another explanation? A child would tell you that
the slinky bounces up and down due to internal energy. Engineers have gone
further. According to engineering text books, any object has three types of
energy: potential due to external fields (or external-potential-energy for short),
kinetic and internal energy, which is mostly due to deformations. Therefore the
total energy of a slinky or any other spring is given by:

$$E_{total} = V + K.E. + E_{int} = V + K.E. + k(\Delta l)^2 \qquad (2\text{-}2)$$

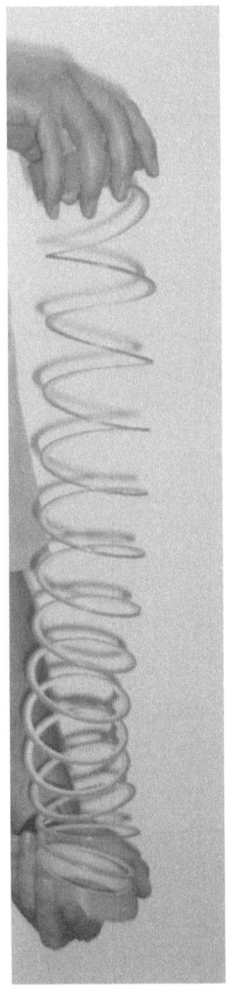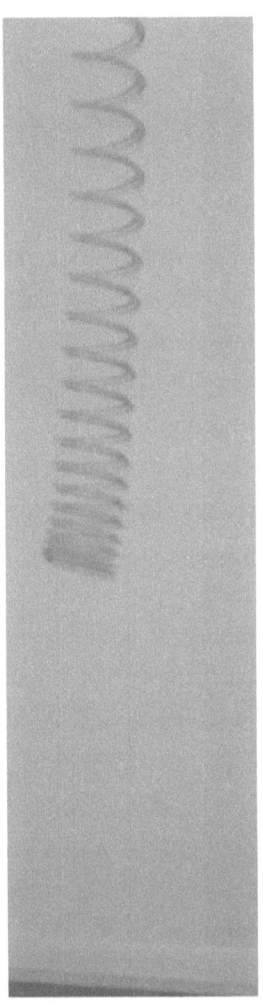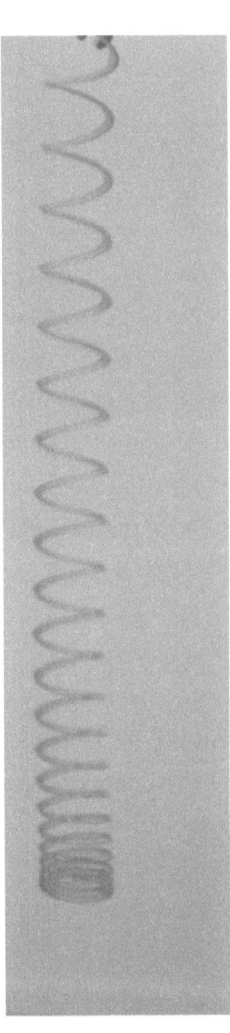

Figure 2-1. An elastic object, for example, a slinky accumulates internal energy due to deformation

Figure 2-2. Minimum slinky extension—virtually no internal energy and high velocity (the coils look blurred)

Figure 2-3. Maximum slinky extension—internal energy peaks and velocity is low

where k is the spring constant and (Δl) is the spring deflection. Spring deflection is proportional to the force that causes the deflection because springs obey Hooke's law. Virtually all objects obey Hooke's law up to a certain deflection, i.e., are elastic up to a certain degree. The total energy of any elastic object is given by:

$$E_{total} = V + K.E. + const \cdot (\sigma)^2 \qquad (2\text{-}3)$$

where σ is the stress. When Hooke's law is applicable, stress is proportional to deflection and equation (2-2) cam be retrieved from equation (2-3).

No object can exist without internal cohesion forces. Therefore, elastic or not, any object has some internal energy. This means, any object has three types of energy. If an object (either macroscopic or sub-atomic) must be studied in detail, equation (2-1) cannot be used.

To rely on equation (2-1) instead of one including internal energy, means to ignore a fundamental law of science: the law of energy conservation. Whoever discards this fundamental law abandons science altogether and embraces mythology wholeheartedly. It is a great mistake to ignore the internal energy of slinky and assume total energy is given by equation (2-1). This mistake leads to some truly spectacular "conclusions". Attach a ribbon to the bottom coil of the slinky. As discussed above, the slinky has no potential and no kinetic energy before release. Therefore, according to equation (2-1) the total energy of the slinky is zero before release. After release, without any energy input, the slinky coils bounce up and down rapidly. As a result the position of a coil or ribbon becomes hard to pinpoint. The slinky appears smeared out over a large region of space—seems to be in many places at the same time (especially when seen through the myopic eyes of ivory tower residents). Hence, some very interesting claims: the motion is illusory, the slinky has no energy and the concept of definite position and velocity is not applicable to elastic objects. A state vector must be used to express position and velocity in a "scientific" manner. This state vector is determined based on careful observations and correlates the probability of detecting a part of an elastic object with a function, ψ, which is called wave function. The probability of detection is maxim when the function $|\psi|^2$ has a maximum and the probability of detection is negligible when $|\psi|^2$ has a minimum.

The physical nature of this correlation cannot be explained if internal energy is ignored. If internal energy is considered, the explanation becomes obvious. When the coil deflection is virtually null (see Fig. 2-4); internal energy, which is proportional to $|\psi|^2$, is negligible and according to the law of energy conservation kinetic energy is almost maximum, see equation (2-2)—as mentioned external-potential-energy is negligible. Therefore, the ribbon speed is high near A. When the slinky deflection, ψ, peaks (see Fig. 2-5), internal energy peaks and kinetic energy is negligible; therefore the ribbon speed is negligible—compare Fig 2-2 and 2-3. As a result, if one takes slinky pictures at random times, the probability of catching the ribbon near A' is much higher than the probability of catching the ribbon near A.

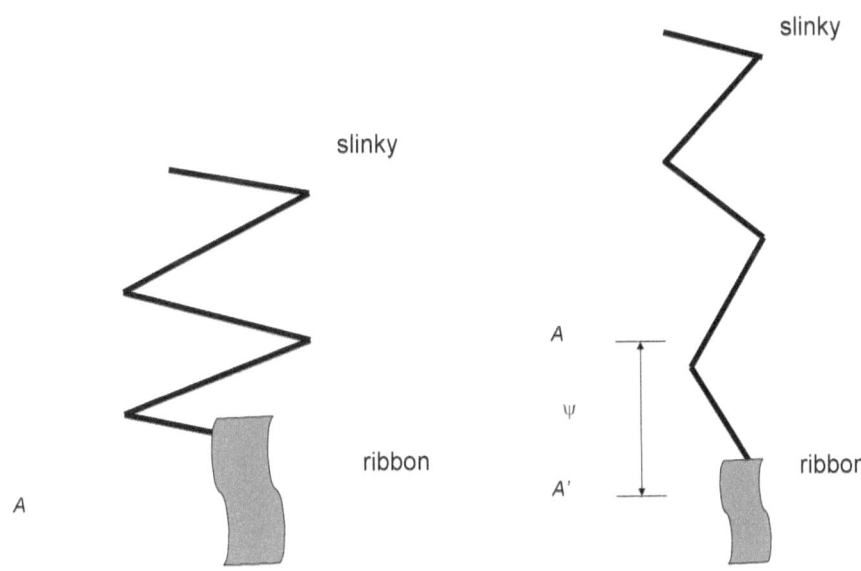

Figure 2-4. ribbon position when slinky deflection is null

Figure 2-5. ribbon position when slinky deflection is almost maxim

In conclusion, after dumping the law of energy conservation, QM physicists had to replace the concepts of well defined position and velocity with a state vector and to claim the dynamics of objects depends on the roll of dice not on deterministic laws. But why stop here? In fact, one must press on. Equation (2-1) cannot be applied without claiming that a slinky is a fundamental building block of matter, an "elementary" particle with no structure. Augmented with this claim, the mythological description of "particle" dynamics becomes even more outrageous: the slinky violates the classic separation between particles and waves; the slinky is a unique object with a dual nature being an "elementary" particle and a wave at the same time.

QM was built on the assumption that "elementary" particles are the ultimate building blocks of matter; therefore all correlations between phenomena taking place within and outside such particles were *a priori* dismissed. This assumption has been invalidated by numerous experiments [1-3]—scattering data proves that "elementary" particles are composite objects with a complex structure. To attempt a study of interactions between such particles ignoring internal phenomena is about as fruitful as trying to understand the causes of the Second World War assuming the political upheavals that took place in Germany, Italy, and Japan had no international consequences. Subsequent refinements of QM did not address this basic oversight.

Internal vibrations, *i.e.* waves, are common in complex structures [4, 5]. In fact, the wave aspect proves an object has a complex structure. This is a matter of elementary physics, yet blithely ignored by quantum physicists for more than 80 years.

Experimental data also prove the components of "elementary" particles are bound by forces approximately proportional to deflection; therefore the total energy of these particles is expressed by an equation similar to (2-2). Nevertheless QM adepts continue to rely on equation (2-1). QM rejects the law of energy conservation. *QM is based on the mythological "law" of energy on-conservation; therefore QM is mythology not science.* If this basic error s corrected, the re-unification of Classical and Quantum Physics becomes a traightforward task. QM is left with no basis and no reason for existence. Correctly applied, Classical Physics provides a deterministic and unitary description of virtually all macroscopic and quantum phenomena (including the quantization of electromagnetic emissions, the double slit experiment, the tunneling effect, radioactivity and the discrete levels of atomic energy). The unexplained phenomena require further studies. Studies based on Classical Physics of course.

2.3. Polarization

A slinky is deformed by pulling the coils apart. But the deformation of "elementary" particles is mainly caused by polarization.

An electric field polarizes a macroscopic object, *i.e.* a system containing many trillions of molecules and atoms, in two ways: by distorting the distribution of electrical charges and inducing a dipole moment or by tending to line up permanent dipole moments that initially had a random distribution. Polarization means deformation. Internal energy accumulates due to deformations. The degree of polarization, *i.e.*, the ratio of internal and external energy, depends on substance density, increasing from less than 0.1% in gases to almost 100% in solids. External and internal energy stored in complex and extremely dense structures such as "elementary" particles are expected to be comparable in magnitude. Both types of energy must be considered when studying the dynamics of "elementary" particles. Gross errors occur if either type of energy is ignored.

The polarization of a macroscopic object can be studied defining an average degree of polarization, but a different model is required when studying a few "elementary" particles. The mutual motion of such particles causes internal oscillations—gradients of external forces tend to distort the internal structure of particles, while cohesion forces tend to restore the structure. Periodic mutual motion causes periodic internal oscillations, which can be represented by means of a Fourier series (see Section 2.4). The internal energy due to polarization adds a time dependent term to the system Hamiltonian. With respect to a fixed frame of reference, the internal oscillations appear as waves that are periodic in both space and time, *i.e.*, as plane waves. These waves are confined within the particle; therefore have a group velocity equal to particle velocity and satisfy Schrödinger's equation. The period of oscillations, T, depends on the particle (or system) Hamiltonian. The internal oscillations are the physical waves sought by Schrödinger until his death. The eigenvalues of internal energy form a discreet set. The law of energy equipartition holds for all oscillators, therefore the ratio of internal and total energy is constant in such systems. As a result, the eignenvalues of internal energy of electrons and the atom eigenstates are proportional.

Due to polarization, the effective electric field fluctuates, *i.e.* depends explicitly on time (see Section 2.4). The effective values of other external fields are similarly affected by internal oscillations. Internal oscillations explain quantum phenomena such as tunneling, radioactivity, and quantization of electromagnetic emissions. Consider for example a particle trapped by a potential barrier. Although the barrier energy exceeds the average kinetic energy, tunneling can occur when most internal energy is converted into kinetic energy. Radioactivity is a similar phenomenon.

The apparent "particle"/wave duality is just an effect of energy conversion, not the defining property of a new class of objects, which are detected only in the quantum realm and have no macroscopic analogue. Spacecraft use internal motion (reaction wheels) to control attitude and these spinning wheels may impart a wave-like aspect to the spacecraft motion. The path of a box containing a heavy mass attached with springs also exhibits this aspect (see Section 3). The lack of any wave aspect would be proof that "elementary" particles are indeed a new class of objects and a new science is required; the apparent duality confirms the universality of Classical Physics.

The assumption that "elementary" particles lack structure led to another error with far reaching consequences—the Coulomb potential is used to quantify the effective electromagnetic potential. As shown by classical electrodynamics, many bodies exhibit some degree of polarization, *i.e.* a reconfiguration of electric charges within a body under the effect of external electric fields. Consequently the dielectric constant of such bodies is not equal to 1, which means the effective field is not equal to the external field. As mentioned, the dielectric constant can be as low as 0.0001 in gases, but up to almost one in solid-state structures—the higher the density of the body, the higher the dielectric constant. Macroscopic bodies, even solids, are collections of atoms situated far apart compared to the atom size. Electrons and nuclei are also relatively far apart. The density of electrons and protons far exceeds the density of macroscopic bodies. Therefore the internal energy stored within "elementary" particles may be comparable with the energy of external fields.

The correlation between internal energy and the dynamics of quantum particles is the key concept underlying the Theory of Internal Energy (TIE), which is presented here. Infinite forces have never been observed. Cohesion forces that bind the components of "elementary" particles are finite. As a result inertia and the gradients of external fields cause internal displacements. Charges located deep inside the particle and charges close to the surface are exposed to different electromagnetic fields—the higher the polarizability of "elementary" particles, the more different these fields. When the external field is constant and the object of study is macroscopic, it can be assumed that each part of the object is polarized to the same degree, *i.e.*, a new stable configuration is reached, and one can define a dielectric constant for the object [6]. For quantum systems like individual atoms, a different approach is required. In such systems, the number of polarized "particles" that are studied is relatively small and therefore each one must be analyzed in detail.

A stable re-configuration requires a perfect balance between external forces and internal forces. On the quantum level, particles are in rapid mutual motion; therefore, a stable configuration is never reached. As a consequence, the components of "elementary" particles are in continuous oscillation.

2.4. Internal Oscillations and Schrodinger's Equation

Internal displacements are periodic; *i.e.*, exactly reproduced after a certain interval [5] if the mutual motion of "elementary" particles is periodic. In such cases, the total energy of a system of particles is conserved.

An arbitrary periodic function can be represented by a Fourier series [7]. As a consequence, the internal oscillations and Schrödinger's wave function satisfy the same equations and have similar properties. QM and TIE rely on almost the same mathematical apparatus. Here is a brief review of this apparatus. In one dimension, the internal displacements (and also Schrödinger's wave) are represented by:

$$\psi(x) = \frac{\sqrt{2\pi}}{L} \sum_{j=-\infty}^{\infty} \exp\left(\frac{2i\pi jx}{L}\right) \varphi\left(\frac{2\pi j}{L}\right)$$

(2-4)

where L is the size of the "elementary" particle. Complex functions are used instead of cosine and sine as a matter of convenience. The size L is very large compared with that of "elementary"-particle-components, and φ is expected to be a continuous function of $k_1 = 2\pi j / L$ (subscript 1 indicates the one dimension). As a consequence, changing n by one unit would have very little impact on each term of the above series; therefore, the sum can be replaced by an integral and equation (2-4) becomes:

$$\psi(x) = \frac{1}{\sqrt{2\pi}} \int_{-\infty}^{\infty} e^{ik_1 x} \varphi(k_1) dk_1$$

(2-5)

Taking into account that $dj = \frac{L}{2\pi} dk_1$ and knowing, ψ, φ (k_1) can be calculated from

$$\varphi(k_1) = \frac{1}{\sqrt{2\pi}} \int_{-\infty}^{\infty} e^{-ik_1 x} \psi(x) dx$$

(2-6)

From equations (2-4) and (2-5):

$$\psi(x) = \frac{1}{2\pi} \int_{-\infty}^{\infty}\int_{-\infty}^{\infty} \exp[ik_1(x - x')]\psi(x')dx' dk_1$$

(2-7)

The propagation of internal oscillations (and of Schrödinger's wave) is determined assuming that at initial time

$$\psi(x) = \frac{1}{\sqrt{2\pi}} \int_{-\infty}^{\infty} [\varphi(k_1)]_{t=0} e^{ik_1 x} dk_1$$

(2-8)

In free space, a wave with propagation vector k_1 oscillates with angular frequency, ω, given by [7]:

$$\omega = \frac{h}{2\pi} \cdot k_1^2 / 2m$$

(2-9)

Therefore, the value of ψ at time t is obtained multiplying each φ (k_1) by

$$\exp\left(-i\frac{h}{2\pi}k_1^2 \frac{t}{2m}\right); i.e.$$

$$\psi(x,t) = \frac{1}{\sqrt{2\pi}} \int_{-\infty}^{\infty} [\varphi(k_1)]_{t=0} \exp\left[i\left(k_1 x - \frac{\left(\frac{h}{2\pi}\right)k_1^2}{2m}t\right)\right]dk_1$$

(2-10)

Equation (2-10) shows the evolution of an arbitrary wave as time passes. From equation (2-10) by differentiating with respect to time:

$$\frac{ih}{2\pi}\frac{\partial \psi(x,t)}{\partial t} = \frac{1}{\sqrt{2\pi}}\int_{-\infty}^{\infty}\varphi(k_1)\frac{\left(\frac{h}{2\pi}\right)^2 k_1^2}{2m}\exp\left[i\left(k_1 x - \frac{\left(\frac{h}{2\pi}\right)k_1^2}{2m}t\right)\right]dk_1$$

(2-11)

Also from equation (2-10):

$$-\frac{\left(\frac{h}{2\pi}\right)^2}{2m}\frac{\partial^2 \psi(x,t)}{\partial x^2} = \frac{1}{\sqrt{2\pi}}\int_{-\infty}^{\infty}\varphi(k_1)\frac{\left(\frac{h}{2\pi}\right)^2 k_1^2}{2m}\exp\left[\left[i\left(k_1 x - \frac{\left(\frac{h}{2\pi}\right)k_1^2}{2m}t\right)\right]\right]dk_1$$

(2-12)

From equations (2-11) and (2-12):

$$\frac{ih}{2\pi}\frac{\partial \psi}{\partial t} = -\frac{\left(\frac{h}{2\pi}\right)^2}{2m}\frac{\partial^2 \psi}{\partial x^2}$$

(2-13)

Assume spin and relativistic effects can be neglected. Then equation (2-13) can be written as [7]:

$$\frac{ih}{2\pi}\frac{\partial \psi}{\partial t} = H(\psi)$$

(2-14)

where H is some function of ψ that does not involve time derivatives of ψ. Hamilton showed that Newton's law of motion can be written as:

$$\frac{d\vec{p}}{dt} = -\nabla\Phi(\vec{x},t)$$

$$(2\text{-}15)$$

$$\frac{d\vec{x}}{dt} = \frac{\vec{p}}{m}$$

$$(2\text{-}16)$$

where Φ is the total potential energy. According to TIE, the total potential energy is given by:

$$\Phi(\vec{x},t) = V(\vec{x}) + E_{int}(t)$$

$$(2\text{-}17)$$

Equation (2-17) shows a key difference between QM and TIE. According to QM, the total potential energy is just V.

Being contained within the particle, the wave defined by equation (2-10) has a group velocity equal to particle velocity. To correctly approach the so-called classical limit [7]:

$$\frac{d\vec{p}}{dt} = -\int \psi^*(\nabla\Phi)\psi d^3x$$

$$(2\text{-}18)$$

and

$$\frac{d\vec{x}}{dt} = \int \psi^* \frac{\vec{p}}{m}\psi d^3x = \int \psi^* \frac{h}{2\pi im}\nabla\psi d^3.$$

$$(2\text{-}19)$$

Therefore, on average, $\vec{p} = -i\frac{h}{2\pi}\nabla\psi$. Based on equations (2-17), (2-18) and (2-19) [7]:

$$\frac{ih}{2\pi}\frac{\partial\psi}{\partial t} = \left[-\frac{\left(\frac{h}{2\pi}\right)^2}{2m}\nabla^2 + \Phi\right]\psi$$

(2-20)

Applying the principle of superposition to a system with n "elementary" particles:

$$\frac{ih}{2\pi}\frac{\partial\psi}{\partial t} = \left[-\frac{\left(\frac{h}{2\pi}\right)^2}{2}\sum_{n1=0}^{n}\frac{\nabla_{n1}^2}{m_{n1}} + \Phi_{sys}(x_1, y_1, z_1, \ldots x_n, y_n, z_n)\right]$$

(2-21)

Up to this point, TIE and QM [7, 8] use the same mathematical apparatus, except for equation (2-17), because both analyze a periodic function linked to "elementary" particles. Internal energy fluctuations average out and both theories predict the same average values. But unlike QM, TIE is fully deterministic, and will provide more accurate predictions for transient phenomena. According to TIE, the Hamiltonian of an "elementary" particle includes a term with explicit time dependence, which satisfies a complex equation:

$$H = V(x) + \frac{mv^2}{2} + ce^{-2i\omega t}$$

(2-22)

where c is a constant and $\omega = 2\pi/T = 2\pi H/h$. Replacing the second and third term on the right hand side with the total kinetic and internal energy, respectively, equation (2-22) becomes applicable to systems with many particles. The Hamiltonian considered by QM proponents, *i.e.*,

$$H = V(x) + \frac{<mv^2>}{2}$$

$$(2\text{-}23)$$

hides the fluctuations of internal energy that cause the apparent particle/wave duality. In addition, equation (2-23) is not suitable for transient phenomena.

In conclusion, the Copenhagen interpretation of quantum physics is based on gross errors. Therefore many claims based on this interpretation, such as the inapplicability of concepts like trajectory or orbit in quantum physics, must be and are rejected here. "Elementary" particles and particle components obey strictly deterministic laws.

The center of mass of a free particle follows a straight path. But the trajectory of particle components, including those detected by instruments, resembles a wave, thus creating the apparent particle/wave duality. Fig. 2-6 illustrates this phenomenon using a simple particle model.

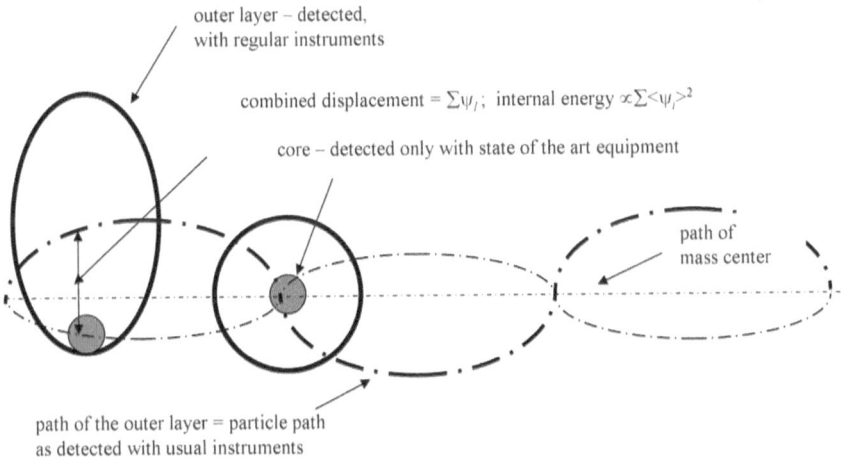

Figure 2-6. The origin of the apparent particle/wave duality illustrated using a simple particle model.

There is a crucial distinction between a structure with internal waves and a free wave. Schrödinger assumed the electron was a free wave. But the analogy with a free-wave is not consistent with experiments—"elementary" particles do not spread out. By definition, internal waves do not spread out either. Internal waves are consistent with all experimental data (see Section 2.7) and with Schrödinger's vision—Schrödinger was never satisfied with a purely probabilistic interpretation and sought a physical wave until his death.

2.5. Probability of Detection and Internal Oscillations

The correlation between probability of detection and internal oscillations is examined in more detail here. Both TIE and QM predict the same correlation rules because both theories use the same mathematical apparatus to predict average values (see Section 2.4). According to both theories; the average momentum and wave function are correlated, $\vec{p} = -i\dfrac{h}{2\pi}\nabla\psi$, etc. But TIE also provides physical explanations. Let us return to the case of a free particle (see also Section 2.2). The law of energy conservation has a notable corollary: if total energy is conserved, potential and kinetic energy are out of phase by 90 deg; *i.e.*, kinetic energy is minim when potential energy is maxim. For simplicity, assume constant y and z velocity and internal oscillations parallel to the x-axis. The probability p(x)dx of detecting a particle between x and x+dx is high when the x-velocity of the particle is low and vice versa. Therefore the probability is high when $|\psi|^2$ is high and low when $|\psi|^2$ is low. See also Section 3. Bound particles obey a similar law. Gravitational potential and kinetic energy can be combined into an equivalent potential [9]. The Coulomb potential and kinetic energy can be combined in a similar manner:

$$V' = V + \frac{1}{2}mv^2$$

$$(2\text{-}24)$$

Therefore, according to Bertrand's theorem, bound electron orbits are possible only in regions where the equivalent potential, V' reaches an extremum [9] and therefore internal energy and ψ also reach an extremum. A stable orbit is possible only if the perturbation force is harmonic (up to a certain deflection, restoring forces are harmonic) or obeys the inverse square law and V' has a minimum. If V' has a minimum, the internal energy and therefore |ψ|², reach a maximum whenever total energy is conserved. The probability current density quantifies the flux of internal energy.

Other charged particles will behave in a similar manner. Other fields, which cause internal deflections, may also trigger internal oscillations. The law of energy equipartition [7 and 10] can be extended to "elementary" particles in energy balance with the surroundings because "... the derivation of this theory involves only the formal properties of the equation of motion. Any other system acting formally like material oscillators must therefore have the same equilibrium distribution of energy." [7]. This means the average internal energy of a free particle must be proportional to the total energy. For particles interacting with external fields, $H = \dfrac{p^2}{2m}$ must be replaced with $H = \sum \dfrac{p^2}{2m} + \Phi$. A similar change is made according to QM [7], except in QM literature V and <p> replace Φ and p, respectively. The average internal and total energy of a system of particles are also proportional. As a consequence, the eigenvalues of both ψ and total energy are discontinuous and ψ provides the energy levels of atoms. Another consequence of energy equipartition is the correlation between particle shape and total energy. High total energy means high internal deflections and therefore distorted shapes.

TIE also explains why electrons may accelerate without emitting electromagnetic radiation. Emission or absorption of electromagnetic radiation occurs only when internal oscillations are in a transient state and internal oscillations remain in steady state while the electron orbit is stable.

The path of "elementary" particles in steady-state motion is similar to a harmonic wave due to two facts: 1) the kinetic energy of a system of point-like masses depends on work exerted by both internal and external forces [11] and 2) the internal oscillations are periodic. The path of an object resembles a harmonic wave, if the object is elastic and has been distorted. Both quantum and macroscopic objects obey this law. Section 3 discusses in detail a double slit experiment [12] carried out with macroscopic objects having a weight of 7.5 kilograms and a speed of more than 2 m/s. The experiment can be performed in a modestly equipped laboratory and shows the wavelengths associated with macroscopic objects having a momentum of more than 15 kg-m/s exceed five centimeters, while de Broglie's equation predicts a wavelength of less than 10^{-31} cm [12].

The correct Hamiltonian, see equation (2-22), predicts the dynamics of quantum systems in a deterministic manner:

$$m_n \cdot a_n = -\nabla \Phi_n (x,t) = -\nabla (V(x) + f(<\psi>_n)) \neq -\nabla V(x) \qquad (2\text{-}25)$$

where $<\psi>_n$ is the average deflection of the nth particle components. Today, the function $f(<\psi>_n)$ can only be approximated because internal forces have not been studied in detail and according to the uncertainty principle, initial displacements cannot be determined. This problem requires more study.

2.6. A Review of Previous Deterministic Models

It is obvious that either "elementary" particles do not obey deterministic laws, or QM fails to take into account some form of energy. For this reason, previous deterministic models sought to augment Coulomb's potential with another form of energy. For example, D. Bohm assumed "elementary" particles are affected by a "quantum potential", U, and created a deterministic model of quantum phenomena on this basis [13]. Bohms's model fails to explain the nature of quantum potential. According to Bohm, the quantum potential is caused by hidden parameters, has wave characteristics and is not observable in macroscopic systems. In reality, quantum potential is distortion energy and all objects, not just quantum particles suffer some distortions. Distortion energy causes internal oscillations. In general, these oscillations are harmonic. Therefore, Bohmian mechanics is equivalent with TIE and therefore correct and QM is incomplete (to say the least). Bohmian mechanics (and hence TIE agree) with all experimental data. Nevertheless, Bohmian mechanics failed to gain widespread acceptance because of reliance on hidden parameters. For QM "physicists", non-deterministic laws of motion, the claim that state vectors provide the most accurate description of reality, the particle/wave duality and cats that are both alive and dead at the same time are ideas that make perfect sense, but Bohmian mechanics does not. QM "physicists" also rejected the idea that some subatomic forces may not be known because quantum particles are not perfectly understood. Ironically, Bohm himself rejected the idea of hidden parameters in a previous work [7].

Another deterministic model was proposed by Nelson [14]. This model is stochastic and also involves hidden phenomena.

The analysis of internal oscillation proves there is no need to invoke new phenomena, not to mention hidden ones. The composite nature of "elementary" particles is a proven fact. The finite value of forces binding the components of "elementary" particles is another proven fact. "Elementary" particles are in mutual motion, which causes an imbalance between external and internal forces, therefore internal oscillations. As a consequence, the Hamiltonian of a system of "elementary" particles must include an additional term with explicit time dependence. If the mutual motion of "elementary" particles is periodic, internal oscillations are also periodic and therefore satisfy Schrödinger's equation. TIE goes beyond reliance on hidden parameters and proves the deterministic nature of quantum laws. In addition, TIE unveils gross errors in the foundation of QM. Because the only alternate explanation must be rejected as erroneous, only the deterministic model of quantum phenomena is acceptable.

2.7. Comparison of QM & TIE Predictions

Any new theory requires experimental validation. Experiments that validate TIE are discussed here. At first glance, TIE is designed to exactly duplicate all QM predictions. Currently, both theories rely on exactly the same mathematical apparatus for predicting the probability of detection and average values. All experiments that validate QM automatically validate TIE. But transient phenomena can be used to determine the correct theory.

Published studies fail to agree on the size of "elementary" particles such as electrons [15, 16]. Furthermore, recent publications indicate that particle size depends on environment [17]. The creators of QM did not even conceive such dependence. But in the context of internal oscillations, this dependence is a simple corollary.

Numerous experiments show that "elementary" particles disintegrate. QM was built on the assumption that these particles are indivisible being the ultimate building blocks of matter. According to TIE, these particles are expected to break up due to collisions and resonance. Resonance increases the amplitude of internal oscillations indefinitely and can destroy any structure either macroscopic of microscopic.

According to QM, "elementary" particles have a dual nature, being both indivisible particles and waves. This duality is inherent. According to TIE, a "particle" showing such duality must be a complex structure. Scattering experiments have confirmed this TIE prediction—"elementary" particles have an internal structure. Some physicists predict that quarks also have an internal structure [18]. According to TIE, the wave aspect of "elementary" particles is just a reflection of internal oscillations, not an inherent attribute. In principle, techniques akin to active noise cancellation could dampen such oscillations below detectable levels, allowing "elementary" particles to move exactly like "classical" particles.

If the total energy of electrons is conserved, an increased level of internal oscillations means less kinetic energy, therefore higher electrical resistance. Dampening techniques could allow breakthroughs in superconductivity research. Laser design and nano-technology may also benefit. Such techniques could also simplify the synthesis of various chemical compounds because valence energy depends on internal oscillations. Materials with dampened internal oscillations might gain additional strength.

TIE predicts a correlation between the probability of finding an "elementary" particle at a given location and the structure of the particle. This prediction could be the basis for another validating experiment (QM predicts nothing of this sort). According to TIE, the probability of finding an "elementary" particle is high if internal displacements are high. High displacements mean high distortion and an asymmetric structure; see Fig. 2-7. Low probability should be associated with low distortion and a roughly symmetric structure.

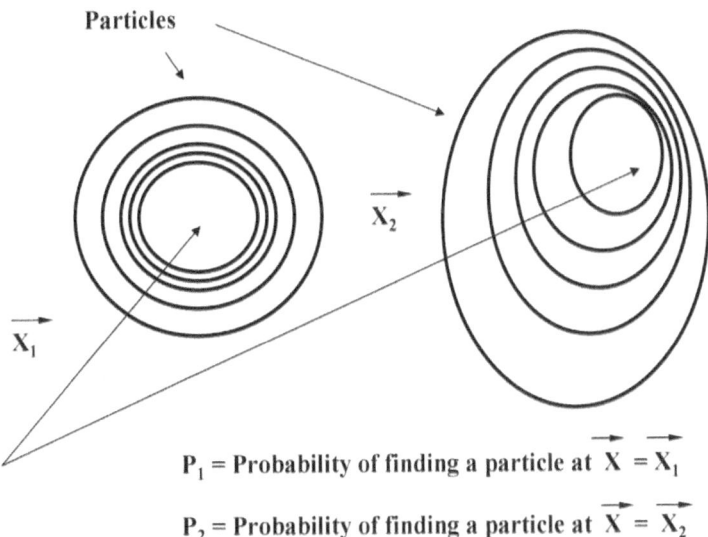

P_1 = Probability of finding a particle at $\vec{X} = \vec{X_1}$

P_2 = Probability of finding a particle at $\vec{X} = \vec{X_2}$

Figure 2-7. The correlation between the probability of detection and the structure of an "elementary" particle.

According to QM, quantum entanglements imply instantaneous action at a distance ("spooky" action, as Einstein said). According to TIE, this phenomenon is just an example of momentum conservation (especially angular momentum).

According to TIE, the interference pattern revealed by the double slit experiment is correlated with internal oscillations. Each interference peak is associated with a deformation extremum. Counting the number of interference peaks, physicists can determine the number of main components of an "elementary" particle.

According to TIE, forces that bind quarks increase with distance. By design, QM is unable to make any prediction about forces that bind the components of "elementary" particles.

2.8. The limitations of QM

Consider a function f that depends on n arguments, $\xi_1, \xi_2, \ldots \xi_n$. If one intends to carry out a scientific study of f, no argument can be neglected a priori. Only a comprehensive sensitivity analysis can indicate which arguments are negligible. Neither ambient nor internal factors can be dismissed a priori. Sometimes, it is possible to ignore an argument a priori and nevertheless predict some correct results, for example average values, but this approach is not reliable and leads to errors in numerous other instances. QM ignores internal energy. Therefore, QM is not rigorous and can provide only a fuzzy description of reality. This clear limitation is well emphasized by the debate centered on the EPR paradox [19].

Here is another example. On average, the kinetic energy of a particle is insufficient to overcome a potential barrier. If the last term in Eq. (2-22) is neglected, the particle appears unable to tunnel out. To provide a QM explanation for tunneling effects, one has to resort to convoluted arguments and numerous assumptions regarding the potential barrier [12]. TIE provides a simple explanation: although the average kinetic energy is too low to overcome the barrier, the peak kinetic energy is sufficient.

If the structure and internal energy of "elementary" particles are considered, Classical Physics provides a flawless description of quantum phenomena, and the need for a new science is eliminated. QM is flawed by design, being a hybrid designed to reconcile physical reality with the gross errors that are embedded in the unalterable core of this mythology.

2.9. Review of Key Concepts

The history of science shows that theories based on fewer and simpler assumptions are likely to provide a more satisfactory description of reality. Up to the interpretation of waves associated with "elementary" particles, QM and TIE rely on common assumptions. These assumptions are discussed in detail in many books and articles, and need no further comment. Of particular interest are the dissimilar assumptions. Table 2-1 summarizes the additional assumptions that lie at the core of QM. The first is completely invalidated by numerous experiments and violates a fundamental law of physics, *i.e.*, the law of energy conservation. The second is also based on the mythical "law" of energy non-conservation. The third and fourth imply a lack of unity in the physical world. TIE requires no peculiar assumptions, and if any assumption included in this new theory is invalidated by experiments, QM is also invalidated.

2.10. In Defense of TIE

As discussed in Sections 2.4, 2.6 and 2.7, TIE cannot be attacked based on past experimental data. An attack based on Classical Physics is *a priori* impossible; this new theory is just a straightforward application of Classical Mechanics and Electrodynamics. To base an attack on QM means to rely on gross errors. QM cannot be separated from gross errors—this "science" loses any reason to exist if the internal structure of "elementary" particles and the law of energy conservation are considered. As shown in Sect. 2.9, an attack against any assumption included in TIE is also a direct attack against QM, but not *vice versa*.

Table 2-1. Peculiar QM assumptions

Assumption	Comments
1) "Elementary" particles have no structure, and therefore have no internal energy.	Invalidated by numerous experiments. Experiments show "elementary" particles are complex structures and binding forces are finite. As a consequence, any external field causes some degree of structural distortion. This distortion depends on the mutual position of particles, therefore changes rapidly and the Hamiltonian of a system of "elementary" particles must include a potential term with explicit time dependence.
2) "Elementary" particles violate the classical separation between waves and particles, and therefore are a new class of objects with no macroscopic analog.	The double slit and other related experiments, which are claimed as indubitable proof of this assumption, merely confirm that "elementary" particles behave like macroscopic objects. The dynamics of all composite objects depends on both external and internal energy and all known objects are composite. Internal oscillations are ubiquitous and harmonic if total energy is conserved. Such internal oscillations imprint a wave-like pattern on the object path. Only paths without any wave aspect would have proven that "elementary" particles are indeed a new class of objects
3) The physical world is not unitary. The dynamics of macroscopic objects is accurately described by deterministic laws, while quantum objects move according to rolls of dice.	Orthodox attempts to bridge this chasm claim the strictly deterministic behavior of macroscopic particles arises from rolls of dice.
4) Concepts such as orbit, trajectory, instantaneous velocity and acceleration have no real physical meaning. The ultimate and most accurate description of reality is a fuzzy state vector.	Being contrary to macroscopic observations, this concept also implies a chasm between the macroscopic and microscopic phenomena.

Von Neumann's argument, the Kochen-Specker theorem, Bell's inequalities, and other "proofs" that a deterministic model of quantum phenomena is inconceivable rely on the assumption that hidden variables are constant parameters and therefore have no bearing on TIE because internal displacements change continuously. In short, no viable strategy of attack is conceivable. But go ahead and try—make my day.

2.11. Conclusions and Recommendations

QM and TIE share some common assumptions, but the former requires four additional ones. The first particular assumption required by QM has been invalidated by numerous experiments. The second one shows complete disregard for a fundamental law of any branch of physics; *i.e.*, the law of energy conservation. The last two imply that macroscopic and microscopic objects must be treated in a completely different manner. Therefore, QM is founded on and forever linked to gross errors. To remove these errors means to remove the need for QM. Applied correctly, Classical Physics provides a flawless and unified description of both macroscopic and quantum phenomena. No new, hidden forces need to be invoked to augment Classical Physics. The particle/wave duality, inferred from the interference pattern revealed by double slit experiments, the correlation between wave function and probability of detecting a particle, and many other quantum phenomena are simply explained by Classical Physics. The others, such as spin require further investigations.

Experiments showing that "elementary" particles might behave like "classical" objects would be a strong argument in favor of TIE. In addition, such experiments are expected to lead to breakthroughs in lasers, superconductivity and nano-technology. QM does not predict the size of "elementary" particles may change. Such size variations support TIE. This new theory also predicts the break-up of "elementary" particles and a correlation between particle structure and the probability of detection.

Von Nuemann, Bell, Kochen, Specker, *et al*, did not prove that a deterministic model of quantum phenomena is impossible. These physicists failed to grasp a simple fact: no object is truly at rest; therefore it is absurd to assume that hidden variables should be constant.

References

1. M. Erdmann, "Proton and Photon Structure", Proceedings of the XXth International Symposium on Lepton and Photon Interactions at High Energies, Rome, Italy, July 2001.

2. H. E. Montgomery, "Proton Structure in Proton-Antiproton Collisions", Fermilab Conference 00/156-E, July 2000.

3. W. Slomirski & J. Szwed, "Remarks on the Electron Structure Function", Proceedings of the IVth Rencontres du Vietnam, Hanoi, Vietnam, 19–25, July 2000.

4. Gh. Buzdugan, "Rezistenta Materialelor", Editia X revizuita, Editura Tehnica, Bucuresti, 1974, pp. 445–446.

5. C. M. Harris, "Shock and Vibration Handbook", 4th edition, McGraw-Hill, 1996, pp. 1.1–1.4, 7.1–7.50.

6. J. D. Jackson, "Classical Electrodynamics", 2nd edition, John Wiley and Sons, 1975, pp. 155–156.

7. D. Bohm, "Quantum Theory", Dover Publications, 1989, pp. 10, 29, 68–69, 78–80, 174, 191–210, 622 and 623.

8. E. Schrödinger, "Quantisierung als Eigenwertproblem I–IV", Annalen.der.Physik 79, 1926, pp. 361–376, 489–527; 80, 437–490; 81, pp. 109–139.

9. H. Goldstein, "Classical Mechanics", 2nd edition, Addison-Wesley Publishing Co., 1980, pp. 76–94.

10. R.A. Strehlow, "Combustion Fundamentals", McGraw-Hill, 1984, pp. 18–20.

11. R. Voinea, V. Ceausu, & D. Voiculescu, "Mecanica", Editura Didactica si Pedagogica, Bucuresti, 1975, pp. 360–362, 500–512.

12. R. Shankar, "Principles of Quantum Mechanics", Plenum Press, 1980, pp. 111–118, 184–185.

13. D. Bohm, "A Suggested Interpretation of the Quantum Theory in Terms of Hidden Variables, I and II", Physical Review 85, 1952, pp. 166–193.

14. E. Nelson, "Derivation of the Schrödinger Equation from Newtonian Mechanics", Physical Review 150 (4), 1966, p. 1079.

15. D.L. Bergman, & J.P. Wesley, "Spinning Charged Ring Model of Electron Yielding Anomalous Magnetic Moment", Galilean Electrodynamics 1, September/October 1990, pp. 63–67.

16. S.J. Brodsky, C.E. Carlson, J.R. Hiller, D.S. Hwang, "Constraints on Proton Structure from Atomic Physics Measurements", SLAC, Publication 10565, WM-04-112, UMN-D-04-5, August 10, 2004.

17. W. Windl & M. S. Daw, "Predictive Process Simulation and *Ab-initio* Calculation of the Physical Volume of Electrons in Silicon", Technical Proceedings of the 2002 International Conference on Computational Nanoscience and Nanotechnology 2002, pp. 197–200.

18. Pati, J. C. and Salam, A., "Lepton number as the fourth "color", Phys. Rev. D10, 275–289, 1974

19. A, Einstein, B. Podolsky, & N. Rosen, "Can Quantum-Mechanical Descriptions of Physical Reality be Considered Complete?", Physical Review 47 (10), May 1935, pp. 777–780.

3

Experiments

The infamous double/slit experiments that supposedly justify the creation of QM prove something else: QM physicists have lost any contact with reality. As a result, all analyses of double slit experiments included in the QM literature are deeply flawed and misleading.

Nomenclature

a amplitude of oscillation
d distance
F force
f function
h Planck's constant
I intensity
k spring constant
k' x-component of wave vector
m crate mass
m' ball mass
p momentum
t time
V velocity
x, y, z Cartesian coordinates
y' mutual distance

Greek

φ phase angle

λ wavelength

ψ wave function

ω angular frequency of internal oscillations

3.1. The Physical Explanation of Double-Slit Results

Waves exhibit a phenomenon called interference, which is specific to them. According to Classical Physics, genuine particles do not exhibit this phenomenon. Classical Physics demonstrates that objects with internal oscillation also exhibit interference, but the founders of QM ignored the structure of "elementary" particles.

Usually, double-slit experiments are performed as follows: let a plane wave, ψ, be normally incident on a screen with slits S_1 and S_2, which are separated by a small distance. Place a row of detectors behind the screen at a distance d. The detectors register the intensity of the wave as a function of position measured along the line AB, see Figure 3-1. If S_1 is open and S_2 is closed, the incident wave will come out of the first slit and propagate outwards. In this case, slit S_1 acts like the virtual source of wave, ψ_1, which passes through and has the same frequency and wavelength as the incident wave, ψ. The detectors register an intensity pattern $I_1 = |\psi_1|^2$. If S_2 is open and S_1 is closed, the second slit acts like the source of a wave, ψ_2, which passes through and also has the same frequency and wavelength as the incident wave, ψ. Wave ψ_2 produces the pattern: $I_2 = |\psi_2|^2$. If both slits are simultaneously opened, both ψ_1 and ψ_2 are registered and the intensity pattern is:

$$I_{1+2} = |\psi_1 + \psi_2|^2 \neq I_1 + I_2$$

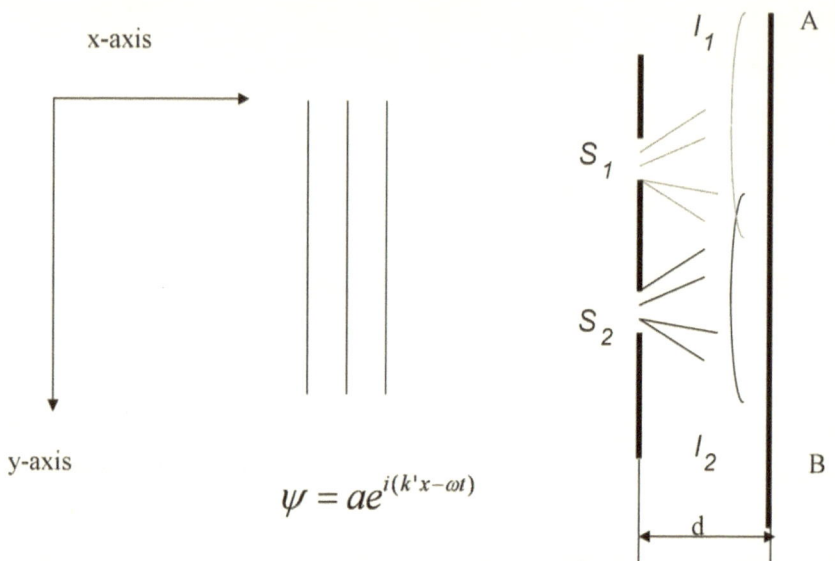

Figure 3-1. Setup for double slit experiments

Genuine (or "classical") particles are supposed to be objects without internal structure and therefore without internal oscillations. This is of course an ideal concept. In reality, all objects have structure and some internal oscillations. But assume by absurd that someone magically creates "classical" particles and repeats the double-slit experiment with such objects. In this case, the source of the incident plane wave is replaced by a source of classical particles, which are sent towards the screen on paths slightly tilted with respect to the horizontal axis (the x-axis) but with the same energy. Place an array of particle detectors on the line AB. By definition, the intensity of a beam passing through the screen, I(y), is the number of particles registered per second at any given coordinate y. The intensity patterns detected when only one slit is open are just like the corresponding patterns formed by waves. But if both slits are kept open classical mechanics predicts that $I_{1+2}=I_1+I_2$ because each particle travels along a definite trajectory that passes via S_1 or S_2 to destination y. If a particle is headed for S_1 it does not matter if S_2 is opened or not. Being ocalized in space, the particle cannot know if S_2 is opened and cannot respond o the opening or closing of the second slit in any way. Therefore the number of particles arriving at x via S_1 is in no way affected by the state of the second lit and vice versa. An objection might be raised: although particles heading oward the upper slit are not aware that the lower slit is open, some of these particles might be deflected by particles coming from the lower slit. The answer to this objection is to send particles one at a time.

The interference patterns obtained with "elementary" particles, even when these particles were sent one at a time resemble those predicted by Classical Physics for waves not those predicted for classical particles. This experimental result was considered definite proof that quantum objects exhibit the infamous particle/wave duality. As discussed above, a classical particle is an ideal concept with no counterpart in reality. Due to internal oscillations, the path of any real object is never a straight line. But one cannot expect ivory-tower residents to bother with reality. QM founders did not even dream to apply Classical Physics to real objects. Because non-existing objects were assumed, Classical Physics predicted an unrealistic result. As software professionals would say, this is a typical example of GIGO (Garbage In, Garbage Out) and cannot be blamed on Classical Physics. Nevertheless, Classical Physics was deemed deficient.

Classical Physics gives accurate predictions for any kind of real objects involved in double-slit experiments. A real object sent toward S_1 may very well arrive at detectors via S_2 due to internal oscillations. For example, assume the double-slit experiment is performed with elastic strings (violin chords for example), see Fig. 3-2. The launch mechanism causes string vibrations (transversal oscillations to be more precise). Therefore these strings are wave quanta. For example, the vertical displacement of a string point is given by (if the string is launched from x=0):

$$\psi = a \cdot \cos(\omega \cdot t + \varphi_0) \cdot \sin(\pi \cdot x / \lambda)$$

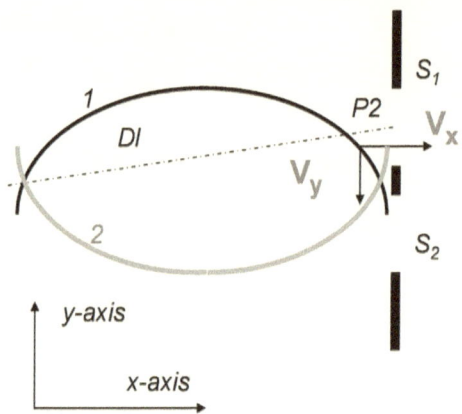

Figure 3-2. Double slit experiment performed with elastic strings

where a and φ_0 are the amplitude and initial phase of oscillation, respectively. λ is the string wavelength.

This vibration does not extend outside the string; therefore is an internal oscillation. Some strings sent toward slit S_1, may actually arrive at the detector via S_1. But according to Classical Physics, other strings sent toward S_1, (for example string 1) will arrive via S_2, see Fig. 3-3. Due to the initial phase of internal oscillation, although the centre of mass of string 1 moves along the dotted line, D_1, towards the upper slit, the leading edge of string 1 is aligned with S_2 when the string reaches the screen. To understand this behavior, consider a point P_2, located somewhere on the string. This point has a horizontal velocity, Vx, imparted by the launch mechanism. The same mechanism also imparts a vertical velocity, Vy(0). But the total vertical velocity at a later time t, Vy(t) is different:

$$Vy(t) = Vy(0) + \frac{d\psi}{dt} = Vy(0) - a \cdot \omega \cdot \sin(\omega \cdot t + \varphi_0) \cdot \sin(\pi \cdot x / \lambda) < 0$$

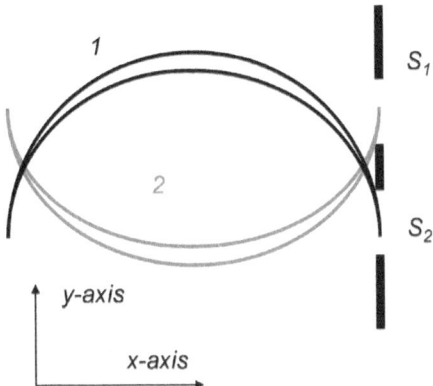

Figure 3-3. Double slit experiment performed with elastic membranes

As a result, point P_2 moves downwards and is exactly aligned with S_2 when this point reaches the screen. Therefore string 1 slithers through slit S_2 and behaves exactly like a wave quanta originating from the lower slit. In a similar fashion; string 2, which was shot along a path tilted downward, slithers through the upper slit and behaves exactly like a wave quanta originating from S_1. Therefore, Classical Physics predicts that when both slits are open, the measured intensity of a beam formed by strings is:

$$I_{1+2} = |\psi_1 + \psi_2|^2 \neq I_1 + I_2$$

Let us repeat the double-slit experiment with membranes, i.e., 2D objects. Just like the strings discussed above, the membranes vibrate. The vertical displacement of a membrane point is given by:

$$\psi = a \cdot \cos(\omega \cdot t + \varphi_0) \cdot f(x, z)$$

where f is a Bessel function, for example, if the membranes are circular. Some membranes sent toward slit S_1, may actually arrive at the detector via S_1. But others, (for example membrane 1) will arrive via S_2. Due to the initial phase of internal oscillation, although the centre of mass of membrane 1 moves along a line tilted towards the upper slit, the vertical velocity of a membrane point P at time t, $Vy(t)$ is:

$$Vy(t) = Vy(0) + \frac{d\psi}{dt} = Vy(0) - a \cdot \omega \cdot f(x,z) \cdot \sin(\omega \cdot t + \varphi_0) < 0$$

As a result, membrane elements move downwards and are aligned with S_2 when reaching the screen. Therefore, membrane 1 slithers through and behaves exactly like a wave quanta originating from the lower slit. In a similar fashion membrane 2, which was shot along a path tilted downward, slithers through the upper slit and behaves exactly like a wave quanta originating from S_1. Again, Classical Physics predicts that the intensity detected when both slits are open is:

$$I_{1+2} = |\psi_1 + \psi_2|^2 \neq I_1 + I_2$$

The "particle/wave" duality, which supposedly sets "elementary" particles apart, is clearly exhibited by a variety of macroscopic objects, and should be called the apparent "complex structure/wave" duality caused by internal energy. This apparent duality is just a trivial consequence of the law of energy conservation, not the hallmark of a unique class of objects. Internal oscillations may occur in any composite object, regardless of scale. Small internal oscillations are harmonic in general. Because the acceleration of all components depends on both internal and external forces, such oscillations imprint a wave like pattern on the trajectory of any composite object. The double slit experiment, which supposedly demonstrates the unique nature of "elementary" particles; can also be faithfully reproduced with macroscopic 3D objects, having, for example, a mass of 7.5 kilograms and a speed of more than 2 m/s. The wavelength of these objects may exceed 5 or 6 cm, depending on the force applied to induce internal oscillations. This experiment can be performed in a modestly equipped laboratory. Classical physics predicts both the wavelength and interference pattern of these objects with immeasurably small errors. By contrast, the de Broglie's wavelength is 10^{-31} cm or less, which means errors can exceed thirty-one orders of magnitude if internal energy is ignored, see Section 3.2 for details. It should be remembered that QM ignores internal energy by design.

A wooden crate with a metallic ball inside may not be the first model of a free "elementary" particle that comes to mind. But although simple, this model is very suitable for showing the so-called dual nature of "elementary" particles is shared by common objects and for illustrating the correlation between internal energy and the probability of detecting a "particle". In addition, plain wooden crates are easy to manufacture and handle in a modestly equipped laboratory.

The ball is attached with two springs to the inner walls of the crate; see Fig. 3-4, forming a system capable of strong internal oscillations. Both springs should be stiff, having, for example, a constant k of 1,000 N/m.

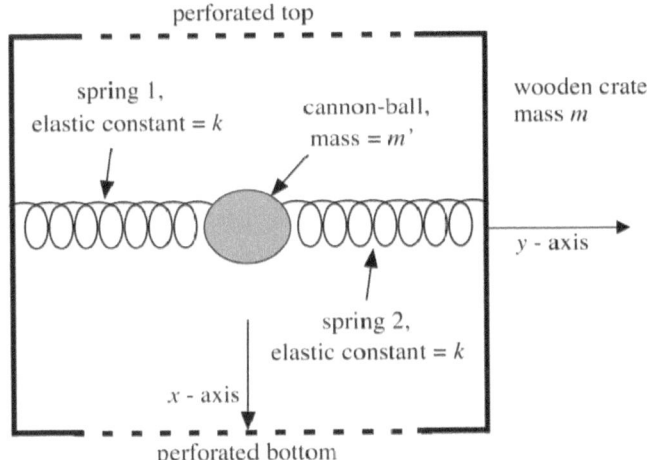

Figure 3-4. A simple model of free particles—a wooden crate with a heavy ball inside

After forcing the ball several centimeters away from the equilibrium position, the crate is dropped. The crate does not fall straight down like a "classical" particle; instead, it visibly oscillates back and forth in the y-direction, which is horizontal. The crate follows a path dictated by both external and internal energy (the crate and ball system stores internal energy in springs). Elastic deformations cause harmonic internal forces, which modulate the path of an object in a harmonic manner. Most common objects exhibit a similar behavior. No matter how smooth the road, a car or truck will not follow a straight line. The engine causes vibrations, which can be easily sensed placing a hand on the hood. To the naked eye, the car path appears to be straight; under the scrutiny of laser instruments the path reveals a wave-like aspect. In short, a "particle" appears "classical" when internal oscillations are undetectable with available instruments.

Choose cubical crates with sides of 0.5 m for easy handling. The crate and ball mass are m=2.5 and m'=5 kg, respectively. The crate is dropped from a height of 0.387 m and reaches a maximum speed of 2.76 m/s. Perforations in the top and bottom reduce the frontal area by 50%, and the drag to weight ratio below 1.6%—a negligible level (measurement errors are expected to be about 5%). Top perforations are also needed to introduce a fork and displace the ball by 5 cm to trigger internal oscillations. The crate is allowed to fall after a delay, Δt, measured from the moment the fork is removed and timed to range from 0 to 0.17 s. The distribution of delays should be uniform. The y-coordinate of the drop point is 0 (the x-axis is vertical and the y-axis is aligned with the direction of oscillations). Therefore, oscillations are not affected by gravity or drag. The crate should be dropped from a platform on rails running parallel to the z-axis. In this manner, the z-coordinate can be changed after each drop, and individual impact points become distinguishable—a spike or protruding nail placed on the bottom of the crate could mark the point of impact on the floor. A vertical drop is the easiest way to launch the crate and is somewhat different than the launch method used in previously described experiments. Therefore the intensity of the interference

attern created by the "crate beam" has a different shape than patterns
discussed above.

Because the crate is bulky, it is not practical to use an actual screen with
slits. Virtual slits can be created by affixing a pointing laser onto a crate side
and placing strips of photoelectric cells on the rear wall of the lab at a height
of 0.100 m; see Fig. 3-5. The crate hits a "virtual" wall when the laser beam
is detected by a photo-cell. In this case, the drop is not taken into account,
which means the impact point should be covered up or flagged.

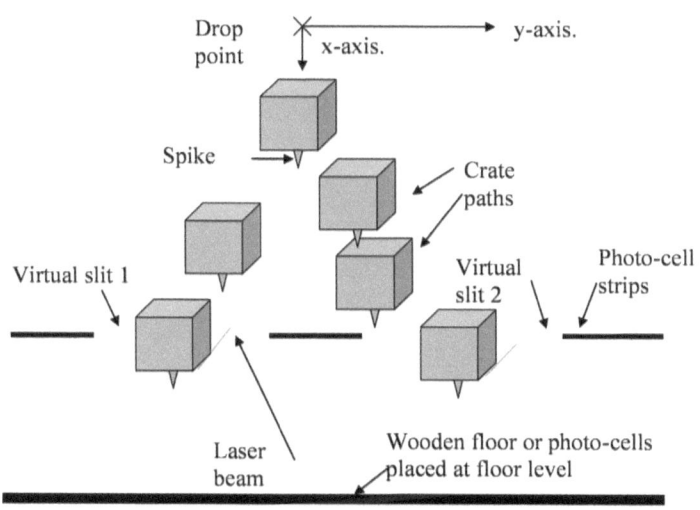

Distance between slits = 0.004 m
Gravity and air drag do not affect crate oscillations.
The crate hits a virtual wall when a photo-cell detects the
laser beam

Figure 3-5. Experimental apparatus for double-slit experiments with 3D objects

A replaceable wooden floor is recommended. Alternately, the impact point
can be determined by additional photocells placed close to the ground. But in
this case, the crate-wave is no longer visible to the naked eye.

3.2. Crate Dynamics

Although an ordinary object, the crate reproduces the behavior of free "elementary" particles. The y-force exerted on the ball is [1]:

$$F_{ball} = m' \ddot{y}_{ball} = 2k(y_{crate} - y_{ball})$$

(3-1)

The y-force exerted on crate is:

$$F_{crate} = m\ddot{y}_{crate} = -2k(y_{crate} - y_{ball})$$

(3-2)

From equations (3-1) and (3-2), taking into account that m'=2m:

$$2m\ddot{y}_{ball} - 2m\ddot{y}_{crate} = (2k + 4k)(y_{crate} - y_{ball})$$

(3-3)

From equation (3-3):

$$\ddot{y}' + \frac{3k}{m} y' = 0$$

(3-4)

where $y' = y_{ball} - y_{crate}$. Equation (3-4) has the solution [2]:

$$y' = ae^{-i\omega t}$$

(3-5)

where a is the oscillation amplitude and ω is the angular frequency of oscillations. By differentiation with respect to time, equation (3-5) yields:

$$i\frac{dy'}{dt} = \omega y'$$

(3-6)

Equation (3-6) is almost identical to Schrödinger's equation; only a constant, *i.e.*, the oscillation frequency, is different. In addition, the oscillation frequency and the wavelength of the crate can be adjusted by changing the spring constant and/or masses. For these reasons, the wooden crate is a suitable model of free-"elementary"-particles.

3.3. Numerical Approach & Analysis of Results

The crate paths were numerically calculated based on equations (3-1) & (3-2) assuming 1700 different time delays, Δt with a uniform distribution. Calculations were performed using MATCHAD 8. These time delays span an interval equal to the period of internal oscillations. By plotting the paths resulting from various time delays, a typical illustration of Feynman's formalism [3] emerges; see Fig. 3-6.

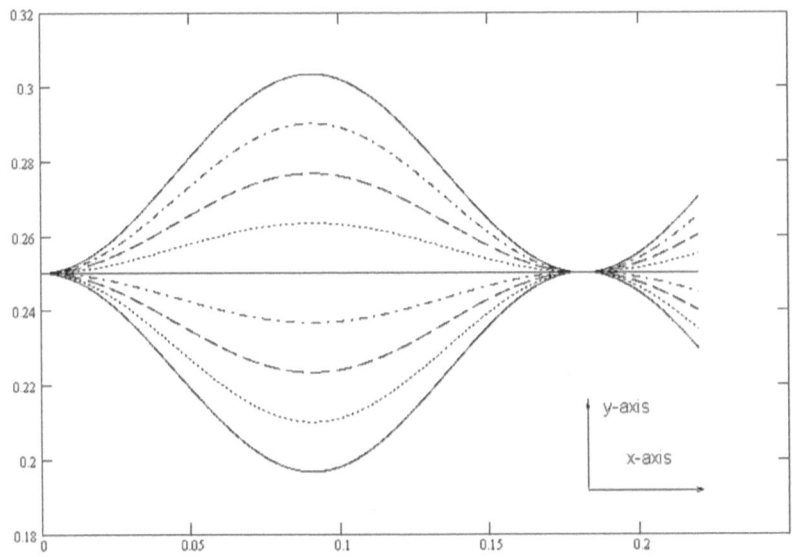

Figure 3-6. Crate paths—an illustration of Feynman's formalism.

Just like "elementary" particles, the crate can pass through either slit, creating an interference pattern, which is visible to the naked eye. The exact pattern depends on the distance between slits, initial ball displacement, masses, drop and slit height, and spring constants. An inter-slit distance of 4 mm yields the interference pattern shown in Fig. 3-7. The impact y-coordinate ranges from 0.0233 m to 0.03 m. Impact points have been grouped in 40 bins. Each bin has a y-size of 0.0016675 m. The number of impact points ranges from just one in bins 1 and 40 to more than 50 in bins 3 and 38, which mark two successive peaks of the "crate-wave". The interference pattern obtained using "classical" particles instead of crates would be very different—all bins would be empty.

Due to internal cohesion forces, any real object is a wave quantum. Therefore, the claim that double slit experiments justify the creation of QM and the rejection of Classical Physics is pure mythology.

As seen from Fig. 3-7, the crate has a wavelength $\lambda \approx 6$ cm. As discussed above, the wavelength depends on ball and crate mass, spring constants and initial displacement. A flexible shell and a larger number of balls and springs would change the wavelength further and would provide a more accurate model of "elementary" particles. But such refinements would exceed the scope of a preliminary study of systems with internal oscillations.

Accurate models, first of "elementary" particles, then of complex atoms and molecules, would be very valuable tools (see Sect. 3.3 for details), but require carefully established criteria for scaling, which are not available today.

The need to study internal phenomena in detail is best illustrated by an example. Shankar [3] calculates the wavelength of macroscopic objects based on de Broglie's formula: "Suppose we do the double-slit experiment with pellets of mass 1 g moving at 1 cm/sec. The wavelength associated with these particles is:

$$\lambda = h / p \approx 10^{-26} \text{cm} \tag{3-7}$$

This value is 10^{-13} times smaller than the radius of the proton."

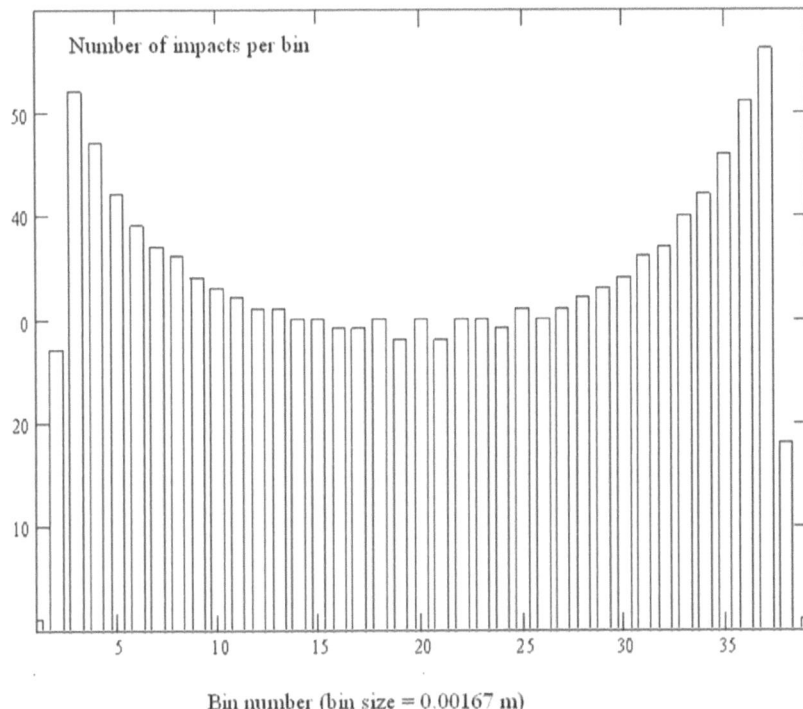

Bin number (bin size = 0.00167 m)

Figure 3-7. The interference pattern observed performing the double slit experiment with crates weighing 7.5 kg—the measured wavelength is 0.0587 meters (y-axis is horizontal).

Based on equation (3-7), objects with a mass of 7.5 kg and a speed of 2 m/s or more should have a wavelength of less than 10^{-31} cm. In this case, customary reliance on equation (3-7) instead of an in-depth study of internal oscillations leads to an error of more than 31 orders of magnitude. Undetected collisions and resonance may augment the oscillations of many "elementary" particles above the level predicted by the Broglie's equation.

Equation (3-5) and (3-6) allow a study of the "crate" wave based on the QM paradigm. The real nature of state vectors is revealed by comparing the classical and QM model of crate dynamics. State vectors are just statistical crutches useful when internal oscillations cannot be studied in detail. State vectors have no direct physical meaning.

3.4. Accurate Models of "Elementary" Particles, Atoms and Molecules

Engineers use scale models to design ships and planes [4 and 5]. Such models have now reached a level of accuracy high enough to render most full-scale tests superfluous, and provide results that cannot be obtained through other means. This development was made possible by accurate scaling criteria based on in-depth understanding of fluid dynamics.

Only the Theory of Internal Energy could allow an in-depth understanding of quantum phenomena, and therefore provide accurate scaling criteria for "elementary" particles, atoms and molecules. Large scale or software models of quantum systems might allow researchers to circumvent Heisenberg's principle of uncertainty [6], and analyze quantum phenomena in unheard of detail.

3.5. Conclusions and Recommendations

The double slit-experiment is supposedly an indubitable proof that quantum particles violate the classical separation between particles and waves, a fact without parallel in the macroscopic realm. Nevertheless, this experiment is faithfully reproduced with ordinary objects such as elastic strings and membranes and a wooden crate with a heavy ball and a couple of springs inside. The equipment required to perform such experiments is simple and available in modestly equipped laboratories. The crate wavelength is visible to the naked eye. Feynmann's path interpretation of QM is also illustrated by experiments with crates. The crate wavelength can be adjusted at will. Therefore, the double-slit experiment so often quoted in quantum mechanics proves only one thing: all quantum physicists have systematically ignored the fundamental law of energy conservation.

Quantum phenomena cannot be rigorously analyzed as long as the tructure and internal oscillations of "elementary" particles are ignored. Only he Theory of Internal Energy accounts for all these factors in a rigorous nanner and is therefore able to provide an accurate description of quantum phenomena. As a consequence, this theory might provide the means to build arge-scale or software models of quantum systems that are accurate enough o circumvent the uncertainty principle and study the dynamics of "elementary" particles in exhaustive detail.

State vectors are just statistical crutches useful when internal oscillations :annot be studied in detail. State vectors have no direct physical meaning.

References

1. C.M. Harris, "Shock and Vibration Handbook", 4th edition, McGraw-Hill, 1996, pp. 2.30–2.33.

2. R. Voinea, V. Ceausu, & D. Voiculescu, "Mecanica", Editura Didactica si Pedagogica Bucuresti, 1975, pp. 500–512.

3. R. Shankar, "Principles of Quantum Mechanics", Plenum Press, 1980, pp. 111–118, 233–241.

4. V. Pimsner, C.A. Vasilescu, & G.A. Radulescu, "Energetica Turbomotoarelor cu Ardere Interna", Editura Academiei Republicii Populare Romane, 1964, pp. 268–270.

5. J.A. Roberson & C.T. Crowe, "Engineering Fluid Mechanics", third edition, Houghton Mifflin Company, 1985, pp. 286–291.

6. D. Bohm, "Quantum Theory", Dover Publications, 1989, pp. 99–115.

4

Internal Energy and Relativistic Phenomena

Section 2 and 3 show the re-unification of quantum and Classical Physics becomes a straightforward task as soon as mythological concepts such as the physical reality of state vectors and particle wave/duality are discarded. These two branches of science are re-unified by a new paradigm based on Classical Physics. The extension of this new paradigm to relativistic phenomena also becomes a straightforward task as soon as two other myths are discarded.

Nomenclature

a constant

b constant

c absolute speed of light in vacuum

E relativistic energy of a free "elementary" particle

E_q energy of an inner quantum

f frequency of quanta emitted by a moving source

f_0 frequency of quanta emitted by a source at absolute rest

h Planck's constant

i index of Cartesian coordinates

L length with respect to moving frame

L_0 length with respect to rest frame

m relativistic mass

m_0 rest mass

m' number of laser pulses required to melt the entire mirror

m" number of laser pulses emitted up to current time

N integer exponent

N_l number of atoms in liquid state

n total number of oscillators within an "elementary" particle

$\vec{n}$ unit vector indicating the path of inner quanta

n_p number of photons in a pulse

R tunnel radius with respect to ship frame

R_0 tunnel radius with respect to tunnel frame

r ship radius with respect to tunnel frame

r_0 ship radius with respect to ship frame

T period of oscillation

t time

t_{cr} critical time

$t_{transit}$ transit time

$\vec{V}$ velocity of "elementary" particle

$\vec{V}_r$ relative velocity

V_r relative speed

V_x, V_y Cartesian components of velocity

V_i i^{th} Cartesian component of velocity

x, y, z Cartesian coordinates

$<\kappa>$ average value of κ

x_{cr} critical distance

Greek:

α angle between the "elementary" particle and quantum velocity

α_s angle between ship and tunnel axis

δ minimum gap between ship and tunnel with respect to ship frame

δ' residual term with negligible value

$\delta^{(m)}$ gap between mirrors with respect to the moving frame

$\delta_0^{(m)}$ gap between mirrors with respect to the mirror frame

γ Lorentz factor

λ wavelength with respect to moving frame

λ_0 wavelength with respect to mirror frame

τ integration variable

4.1. Overview of Relativity Theories

The null results of Michelson and Morley (MM) experiment [1 and 2] means either the speed of light in vacuum is a genuine constant, or a hitherto unknown phenomenon masks the true speed. The classical theorem of velocity addition excludes the first alternative and initially physicists concluded the measured and true speed of light are different. The quest for unknown phenomena responsible for this difference has been thorny. Two schools of thought emerged early on: either the MM experiment was badly flawed and must be dismissed altogether, or the null result is due to an aether effect, this being a hypothetical medium with particular properties (such as all pervasive, akin to both a gas and a solid, etc). For decades, all explanations of relativity based on Classical Physics assumed the existence of aether. The most renowned of these theories was published by Lorentz in 1904 [3]. He assumed relativistic effects are caused by aether pressure. Although rejected later, Lorentz' theory remained important due to two major contributions: proof that length contraction, mass variation, and time dilation are correlated and the Lorentz transformation, which was incorporated by Einstein in the Special Relativity Theory (SRT) [4].

Lorentz's theory was not invalidated by experimental data. But according to Occam's razor, the particular aether hypothesis was considered reason for dismissing all theories based on Classical Physics.

Lorentz showed the speed of light may appear constant with respect to all inertial frames of reference and natural phenomena may appear to run their ourse according to exactly the same general laws with respect to all inertial

rames of reference. For example, assume two spaceships I and II scan one

nother with radar signals, see Fig. 4-1. Assume ship I is at rest and ship II is

noving toward the first with a relative speed, V_r equal to $\dfrac{c\sqrt{3}}{2}$. In this case,

ne length of the second ship shrinks according to the Lorentz transformation:

$$L_{II} = L_0 / \gamma_{II} = L_0 \cdot \sqrt{1 - \frac{V_r^2}{c^2}} = \frac{L_0}{2}$$

$$(4\text{-}1)$$

The length of ship II measured with respect to ship I, $L_I{}^{II}$ is therefore:

$$L'_{II} = L_{II} = L_0 \cdot \sqrt{1 - \frac{V_r^2}{c^2}} = \frac{L_0}{2}$$

$$(4\text{-}2)$$

Ship I

Ship II

Figure 4-1. An illustration of the principle of relativity. Two spaceships scan each other with radar signals and measure the same length contraction and mass increase

The mass, m, of any atom inside ship II increases as follows:

$$m = m_0 \gamma_{II} = \frac{m_0}{\sqrt{1 - \frac{V_r^2}{c^2}}} = \frac{m_0}{2}$$

$$(4\text{-}3)$$

The mass of the ship also changes according to equation (4-3) and the frequency of every electromagnetic signal originating from ship II, including radar signals, changes as follows:

$$f_{II} = f_0 \cdot \sqrt{1 - \frac{V_r^2}{c^2}}$$

(4-4)

Because ship II is a moving source, an observer on-board ship I measures the following blue-shifted frequency:

$$f_{II}^I = \frac{f_{II}}{1 - \frac{V_r}{c}} = \frac{f_0 \cdot \sqrt{1 - \frac{V_r^2}{c^2}}}{1 - \frac{V_r}{c}} = f_0 \sqrt{\frac{1 + \frac{V_r}{c}}{1 - \frac{V_r}{c}}}$$

(4-5)

Radar signals emitted by ship I and reflected by the bow, A' and antenna B' of ship II, return at slightly different times. The delay between these two radar returns is given by:

$$\Delta t = L_{II} / c = \frac{L_0}{c} \cdot \sqrt{1 - \frac{V_r^2}{c^2}}$$

(4-6)

According to Lorentz' theory, physical phenomena (including clocks rates) slow down on board ship II. The result is the so called time-dilation, which is given by:

$$t_{II} = t_0 \cdot \sqrt{1 - \frac{V_r^2}{c^2}}$$

(4-7)

According to the Classic Doppler effect, the frequency of radar signals beamed by ship I with respect to ship II is:

$$f_I' = f_0 \left(1 + \frac{V_r}{c} \right)$$

(4-8)

Equation (4-8) does not take into account time dilation on-board ship II. If time dilation is factored in, the frequency measured on-board ship II is given by

$$f_I'' = \frac{f_I'}{\sqrt{1 - \frac{V_r^2}{c^2}}} = \frac{f_0 \cdot \left(1 + \frac{V_r}{c} \right)}{\sqrt{1 - \frac{V_r^2}{c^2}}} = f_0 \sqrt{\frac{1 + \frac{V_r}{c}}{1 - \frac{V_r}{c}}}$$

(4-9)

Equation (4-5) and (4-9) demonstrate that observers on both ships measure identical blue-shifts. These observers cannot determine which ship is at rest and which one is moving by comparing the measured frequencies of radars signals. Spectral lines emitted by an atom depend on the atomic mass. Because frequency shifts are identical, observers from both ships measure identical shifts in spectral lines emanating from the other ship. As a result both observers may conclude the total mass of the other ship has in increased according to equation (4-3).

As a result of time dilation, the length of ship I with respect to ship II is calculated as follows:

$$L_I'' = c \cdot \Delta t = L_0 \cdot \frac{c}{c} \cdot \sqrt{1 - \frac{V_r^2}{c^2}} = \frac{L_0}{2}$$

(4-10)

Where the delay, Δt is given by equation (4-6).

In conclusion, observers on both ships detect the same length contraction, frequency and mass change. Equation (4-1) through (4-10) seem to validate two celebrated "principles": 1) natural phenomena run their course according to exactly the same general laws with respect to all inertial frames of reference and 2) the speed of light in vacuum is constant with respect to all inertial frames of reference (these two principles cannot be separated). In addition, Occam's razor rules against the aether hypothesis. Therefore why not eliminate this hypothesis and develop a theory of relativity based on the above paired principles? This is the logic that led Einstein to create the Special Relativity Theory (SRT). Because Lorentz transformations are included in SRT, some people, even some physicists assume these two theories are in fact equivalent. This is a catastrophic mistake. There are profound differences between these two theories. Lorentz's theory is based on Newton's paradigm. SRT departs from Newtonian physics in a radical manner. To quote Einstein [4]: "... on the basis of Theory of Relativity the method of interpretation is incomparably more satisfactory. According to this theory, there is no such thing as a 'specially favored' (unique) coordinate system to occasion the introduction of the aether idea, and hence there can be no aether drift, nor any experiment with which to demonstrate it. Here the contraction of moving bodies follows from the two fundamental principles of the theory, without the introduction of particular hypotheses; and as the prime factor involved in this contraction we find, not the motion in itself, to which we cannot attach any meaning, but the motion with respect to the body of reference chosen in the particular case point."

Therefore Einstein claims SRT must supersede Classical Physics because the latter cannot explain relativistic phenomena without the aether hypotheses. Ironically, in order to reject the classic explanation of relativity, Einstein was forced to base SRT on four particular (even peculiar) hypotheses (not principles as usually labeled). Only two are clearly stated in SRT: **1)** "If relative to K, K' is a uniformly moving co-ordinate system devoid of rotation, then natural phenomena run their course with respect to K' according to exactly the same general laws as with respect to K. This statement is called the principle of relativity (in the restricted sense)" [4], and **2)** the speed of light in vacuum is a constant with respect to all inertial frames of reference. The third one is tacit [5]: "A fixed rod that is at rest in the system S and is of length 1 cm, will, of course, also have the length 1 cm, when it is at rest in the system S', provided that the remaining physical conditions are the same in S' as in S. Exactly the same would be postulated of the clocks. We may call this tacit assumption of Einstein's theory the principle of the physical identity of the units of measure".

There is yet another tacit assumption: there is no transversal contraction regardless of velocity. Otherwise, consider the following paradox: a relativistic bullet is approaching a hole in a wall. The rest radius of bullet is slightly smaller than that of the hole. When the bullet motion is studied with respect to the wall frame of reference, the bullet radius appears even smaller; therefore the bullet is expected to pass through. With respect to the bullet frame of reference, the hole radius would appear smaller than the bullet radius and therefore the bullet would not be expected to pass through, *i.e.*, natural phenomena would not run their course according to the same laws of physics with respect to these two Galilean frames of reference. Therefore the first SRT hypothesis would be proven invalid. Lorentz has argued the transverse contraction factor must be almost equal to unity [3], but this would not be enough. As a result, SRT was patched up with the second tacit hypothesis. Amazingly, the patch was accepted for a century. A careful review shows either the omission or the inclusion of this hypothesis destroys SRT; see Section 4.3. Based on Euclidian geometry, which is valid in Galilean frames, if the x-axis of Galilean frame K is parallel with the x'-axis of frame K'; the angle between the relative velocity and the x-axis is equal to the angle between the relative velocity and x'-axis. It is interesting to consider the alternative; *i.e.*, the bullet velocity is aligned with the hole-axis with respect to one frame but not with respect to the other. As a result, the bullet would pass through the hole with respect to the first frame, but not with respect to the second. A more in-depth analysis might reveal other particular hypotheses, but would exceed the scope of this review.

History indicates that superior theories are simpler and rely on fewer hypotheses. For this reason alone, an alternate theory of relativity based on Classical Mechanics, which avoids the aether hypothesis, would seriously challenge SRT. Unfortunately, previous attempts to develop such a theory were plagued by other particular hypotheses. For example, to prove the magnetic field generated by a moving charge is the cause of relativistic phenomena, Marmet [6] assumed the Biot-Savart law, which was derived for very numerous charges can be extended to a single electron, the electric and magnetic fields surrounding a single electron are isotropic and mass is just the energy of electric fields. In order to extend the explanation to neutral particles, Marmet also assumed that absolutely all particles have components with electric charges and "the apparent absence of external electric field around neutral particles, is explained by the presence of a very large number of small electric dipoles, formed by small electric bubbles of positive and negative fields", *etc*. Nevertheless, the simplest theory of relativity, which agrees with all experimental data and includes no particular hypotheses, is based on Classical Physics, see Section 4.7.

4.2. The Art of Hypothesis Selection

Inconsistent theories are shortly rejected, and merit no further mention. The ultimate fate of a consistent theory does not depend on the erudition of the author, the demonstration clarity or the elegance of the mathematical apparatus involved. The ultimate fate depends only on the quality of underlying hypotheses. After taking a wrong turn at an intersection, even the best car in the world would not take a stubborn driver to the proper destination. Many times ambiguous experimental evidence has masked the correct path, and science has turned onto dead-end roads. Equal consideration must be given to all alternate hypotheses. All possible theories based on these hypotheses should be fully developed and exhaustively tested until all but one are invalidated by experiments or contradictory predictions. But, all too often official "science" wholeheartedly embraces one hypothesis and the resulting theory, no matter how absurd, is raised to the level of sacred dogma. As a result, the few scientists willing to pursue the truth no matter what the consequences are forced to wage an uphill battle and the progress of science slows to a crawl.

For example, the perceived rotation of all celestial bodies around Earth left astronomy at a crossroad. The Heliocentric explanation was first considered in antiquity, but was rejected because the geocentric alternative seemed obvious. This choice led to Plato's astronomy a grotesque caricature of science, which strangled scientific progress for more than fifteen hundred years and is an epitome of the Dark Ages. Beware of obvious hypotheses!

Winston Churchill said "the only thing one may learn from history is that nobody learns anything from history". At the beginning of the twentieth century, when the discovery of the wave aspect of "elementary" particles left science at another crossroad, physicists jumped to the obvious explanation, concluding the particle-wave duality is genuine and "elementary" particles are objects with unique properties (see Section 2). In the meantime, numerous experiments have revealed the complex structure of "elementary" particles. Any engineering student knows most structures are elastic to some degree, and therefore the structure components may be involved in harmonic oscillations. As a result, the dynamics of these structures cannot be properly modeled if the energy of oscillations, *i.e.*, internal energy, is neglected a priori. Because Nature is unitary, the dynamics of "elementary" particles depends on internal energy like the dynamics of any macroscopic structure (see Section 2 and 3). This fact means the particle-wave duality is not genuine and to neglect internal energy means to ignore the law of energy conservation. Nevertheless, Bohr, Dirac, Heisenberg and all the other adepts of the Copenhagen school of Quantum Mechanics ignored this most basic law of physics for the sake of an "obvious fact", which in reality is just an illusion, proving again Churchill's wisdom. As a result, Quantum Mechanics, the "science" based on the "obviously" dual nature of "elementary" particles, rivals the absurdity of Plato's astronomy. D. Bohm, has correctly advocated the alternate explanation, *i.e.*, the duality is apparent, not genuine, and is caused by an additional force having a value that varies in a harmonic manner [7]. But Bohm did not understand the origin of this force. Bohm assumed hidden phenomena of unknown type are the source of this force. As a result, Bohmian mechanics did not gain wide acceptance. Once again reliance on a particular hypothesis caused an impasse.

The Michelson-Morley experiment placed science at yet another crossroad. As mentioned above, there are two possible explanations: the speed of light in vacuum is a genuine constant or appears to be constant due to some overlooked phenomena. Lorentz selected the second explanation but used a questionable hypothesis to justify his choice. As a result, although in agreement with all experimental data, Lorentz' theory was rejected. Einstein assumed the speed of light is a genuine constant (as peculiar a hypothesis as any, and not a principle as labeled) and created a theory full of intractable contradictions not paradoxes as previously claimed. Theories based on particular hypotheses cannot be always avoided, but should never be trusted. Such theories are nothing more than stepping stones, and should be discarded as soon as possible. Nevertheless, official "science" raised Einstein's relativity to the level of sacred dogma.

4.3. Relativistic Paradoxes and the Absolute Frame of Reference

Herbert Dingle has discussed an interesting paradox [8]. Consider two observers A and B at rest in two inertial frames, K_a and K_b, respectively. The relative velocity of these frames is high enough to cause observable relativistic effects. To prove relativity, Einstein compares time readings and rates of clocks placed in different frames of reference [4]. Therefore based on Einstein's teachings; observers A and B can and should compare clock rates.

Observer A can send a light beam toward observer B whenever clock A indicates 0 seconds, while observer B can send a beam toward observer A whenever clock B indicates 0 seconds. In this manner, both observers can determine which clock runs faster and can compare notes by radio. Based on SRT both observers expect the clock of the other to run slower, but the exchange of notes will prove one expected result is wrong. Or is there a third possibility? According to some SRT advocates, there is no contradiction in Dingle's paradox, just a confusion caused by an inappropriate choice of words because there is no such thing as the "moving clock". Each observer is equally entitled to regard the other as moving. Problem solved right? Not quite. If clock A runs slower than clock B with respect to one frame but not with respect to the other the first SRT "principle" is invalidated; see also the next paradox. Therefore either SRT provides erroneous predictions or is based on an untenable assumption.

Lorentz' theory predicts the clock of the observer having the smallest absolute velocity (*i.e.*, velocity with respect to the absolute frame of reference) runs faster. Therefore Lorentz' theory does not lead to an intractable contradiction.

Feeling entitled to a personal opinion may lead to more serious consequences than a disagreement about clock rates, sometimes downright catastrophic. Consider another well-known paradox. A starship traveling at relativistic speed passes through a long space tunnel. Frames K_t and K_s are linked to the tunnel and starship respectively. With respect to frame K_t, the starship length is negligible compared to the tunnel length and *vice versa* with respect to K_s. As a result, an observer A at rest with respect to frame K_t will see the starship enter the tunnel; disappear from sight for a time and then exit. With respect to frame K_s, the ship will never be completely inside the tunnel. With respect to this frame, the bow emerges from the exit before the stern enters the tunnel. SRT advocates treat this paradox as an amusing difference in perspective caused by the relativity of simultaneity, which means that events do not occur in the same sequence with respect to different frames of reference.

But a real tunnel transit would be more complicated because nothing is perfect. Therefore one should expect a misalignment between the starship and tunnel axis. Thus, the ship velocity will have a small component normal to the tunnel axis, V_y in Fig. 4-2. The contradiction arising in this context cannot be explained invoking the relativity of simultaneity because the outcome does not depend on simultaneity. A concrete example will clarify the problems caused by the slightest misalignment. Assume the relative speed of ship with respect to tunnel, V_r, is 0.9999875c and the angle between the ship and tunnel axis, α_s, is 0.00057296 radians, which means the x-axis and y-axis components of velocity (V_x and V_y) are 299,996,235 m/s and 3000.1 m/s respectively (c has been rounded up to 3×10^8 m/s). With respect to frame K_t, the Lorentz' factor, $\gamma = 1/\sqrt{1 - V^2/c^2} = 200$, ship radius is r = 10 m and tunnel radius is $R_0 = 20$ m (the subscript 0 indicate values with respect to the rest frame). Therefore the gap between the ship and tunnel wall is 10 m. The time required to close this gap with respect to frame K_t, is $t_{cr} = 10/3000.1 = 0.0003333$ sec. During this period of time, the ship will travel an axial distance $x_{cr} = 999,542$ m $< L_0$.

As a result, with respect to frame K_t, natural phenomena run their course according to the laws of catastrophic matter-antimatter annihilation, because halfway through the tunnel, the ship impacts the wall with relativistic speed. Due to this collision, the fields containing antimatter fuel collapse and the entire amount of fuel (having a rest mass of several million tons and therefore a relativistic mass of almost a billion tons) is instantaneously and completely converted into energy. The resulting explosion rivals a supernova, everything turns into dust up to a distance of several light hours and all colonies located in nearby star systems are destroyed. Frame K_t does not remain Gaussian (not to mention Galilean) during the explosion, but this change does not prevent the catastrophe, just complicates the prediction of explosion details.

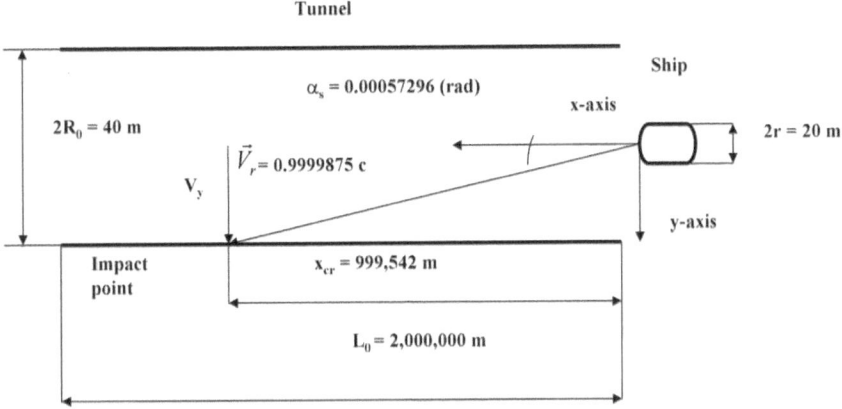

Figure 4-2. Tunnel and ship paradox with respect to the tunnel frame of reference, K_t.

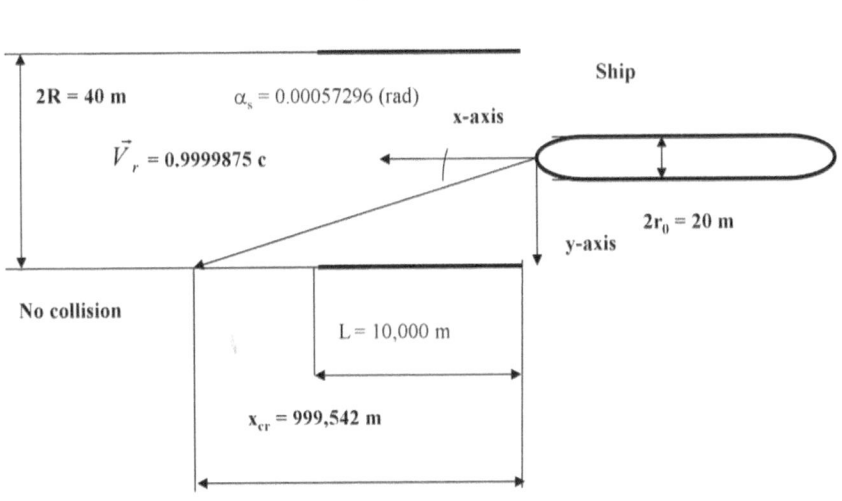

Figure 4-3. Tunnel and ship paradox with respect to the ship frame of reference, K_g.

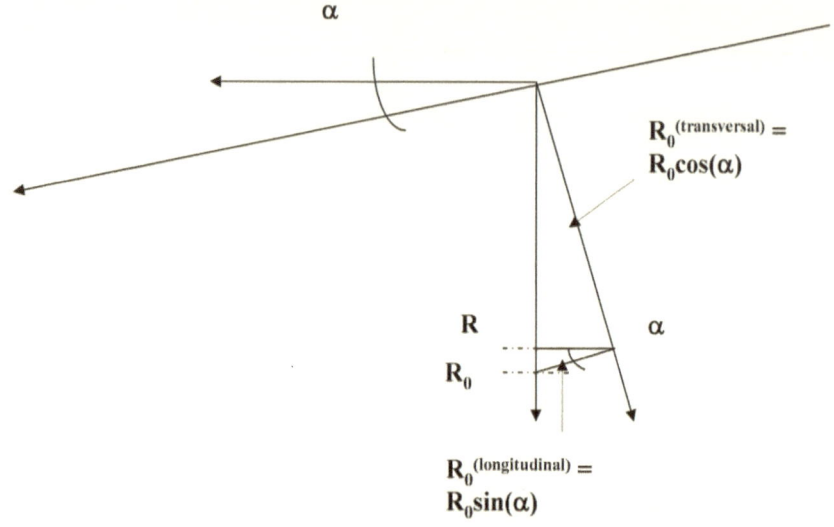

Figure 4-4. Tunnel radius measured with respect to starship frame, K_s

With respect to frame K_s, see Fig. 4-3, natural phenomena run their course according to other laws, because the tunnel length shrinks from $L_0 = 2,000,000$ m to $L = 10,000$ m, while the tunnel radius remains virtually equal to R_0. As seen in Fig. 4-4, the tunnel radius can be projected on the ship axis and the direction normal to this axis. The first component is small to begin with, $i.e.$, $20\sin(\alpha_s) = 0.0002\,\text{m}$, and shrinks to only 0.000001 m due to longitudinal contraction. The normal component is $20\cos(\alpha_s) = 19.999999999\,\text{m}$, and is the same with respect to both frames of reference because the transverse contraction factor is supposed to be exactly 1 (see the SRT patch discussed in Section 4.1.). As a result, the tunnel radius with respect to frame K_s is

$$R = \sqrt{0.000001^2 + 19.999999999^2}$$, $i.e.$, virtually equal to R_0 leaving plenty of room for a safe ship passage (without the patch discussed above, SRT could predict collision with respect to both frames). The ship diameter is also virtually the same with respect to both frames. The x and y components of velocity are

:he same with respect to both frames (see Section 4.1.). This means that with respect to frame, K_s, the time of transit through the tunnel will be

$$t_{transit} = L/V_y = 10,000/299,996.235 = 0.000033334 \text{ (sec)}$$

and the minimum gap δ, will be:

$$\delta = R_0 - r_0 - V_y \cdot t_{transit} = 9.9 \text{ (m)}$$

Therefore SRT predicts the downward drift is negligible with respect to frame K_s, and the transit will be perfectly safe, which also means frame K_s remains Galilean at all times (yet another difference between predictions). The ship crew has every reason to enter the tunnel because even in ideal circumstances, a ship re-alignment is a delicate and lengthy operation. When circumstances are less than ideal or time is running out, re-alignment is out of the question. Aborting the passage means plunging the ship into a cosmic region crisscrossed by radiation so lethal the traffic control authority had no choice but to build a protective tunnel costing countless trillions, not to mention a large number of support facilities. But first, the ship has to veer away from the tunnel without turning all passengers and crew-members into a thin film of organic matter coating all inner ship surfaces, a very improbable maneuver, given the short time and relativistic speed involved. Based on SRT, the tunnel crew expects a catastrophe if the ship attempts to pass through the tunnel, while the ship crew expects a catastrophe if the ship does not attempt to pass through the tunnel.

Therefore, none of the people involved will praise the argument that the tunnel and ship crew are entitled to contrary opinions. The argument is also counter-productive. There are only three possible outcomes if the ship attempts to pass through the tunnel: 1) an uneventful passages with respect to both tunnel and ship frame, 2) a catastrophe with respect to both frames, 3) an uneventful passage with respect to one frame and a catastrophe with respect to the other. Outcome 1 and 2 mean SRT predictions are wrong, and invalidate the argument that the crews are entitled to contrary opinions. As far as SRT is concerned, outcome 3 is even worse. Outcome 3 means that although relative to K_t, K_s is a uniformly moving co-ordinate system devoid of rotation, natural phenomena ***do not run*** their course with respect to K_t according to exactly the same general laws as with respect to K_s (for example the laws of matter-antimatter annihilation play no role with respect to frame K_s because the main ship engines are turned off), which means the first SRT hypothesis is invalid.

According to Lorentz' theory only one outcome is possible: a catastrophe caused by the ship/tunnel collision, because the ship contraction is real, while the tunnel contraction with respect to frame Ks is an illusion caused by the relativistic speed of this starship. As before, Lorentz' theory does not lead to an intractable contradiction.

Because axial and transversal contractions are different, numerous natural phenomena must follow a different course with respect to different Galilean or Gaussian frames. Consider drag for example. The density of gas depends on gas mass and volume therefore will not be the same with respect to a frame linked to a high speed probe and to a frame linked to a volume of gas. The drag coefficient depends on the probe shape, which is spherical with respect to the first frame and ellipsoidal with respect to the other. As a result, the drag coefficient calculated with respect to one frame is not equal to the drag coefficient calculated with respect to the other. Material properties such as strength and thermal conductivity would also differ. Because these changes are not necessarily balanced, SRT would predict the probe passage is uneventful with respect to one frame, but causes the probe destruction with respect to the other. As usual, Lorentz' theory predicts the same outcome with respect to both frames, therefore does not lead to an intractable contradiction.

Objects that are on a collision course with respect to one frame do not collide with respect to the other, a situation that occurs frequently on quantum level. Two atoms may be on a collision course and about to undergo fusion with respect to one frame, but not with respect to another. With respect to one frame, the course of natural phenomena leads to fusion and conversion of mass into energy. With respect to another frame, fusion does not occur. No wonder that previous attempts to reconcile Quantum Mechanics and relativity did not succeed (see also Section 4.4).

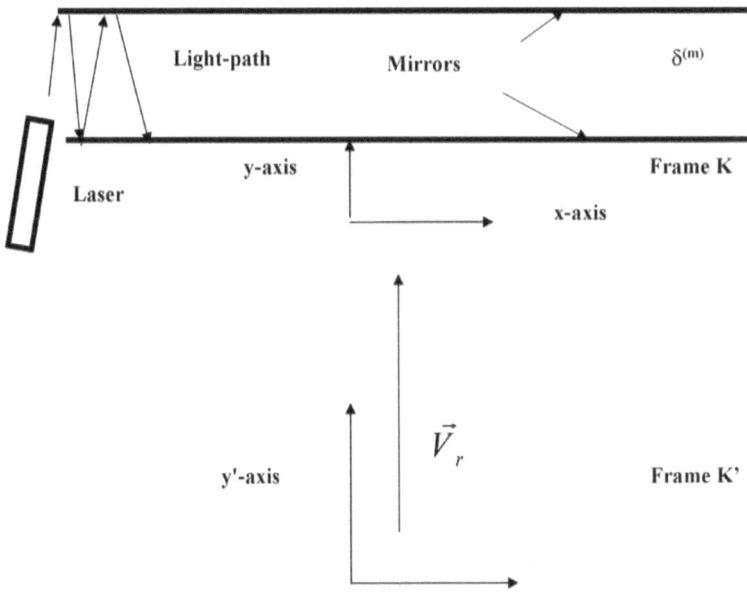

Figure 4-5. Steady-state electromagnetic waves would exist with respect to frame K but not with respect to frame K'

Geometric distortions also lead to contradictions in scenarios involving radiation. Consider light pulses emitted by a laser, which are then reflected by parallel mirrors having a reflection coefficient virtually equal to 1 and separated by a small gap, $\delta^{(m)}$, see Fig. 4-5. A laser pulse contains n_p photons. With respect to Galilean frame K, where the mirrors and laser are at rest, the light wavelength is λ_0 and the gap between mirrors, $\delta_0^{(m)}$ is exactly equal to $\lambda_0/2$; therefore wide enough to prevent mirror atoms from absorbing these photons (mirror atoms absorb and convert electromagnetic radiation into heat whenever $\delta_0^{(m)} < \lambda_0/2$). Therefore, with respect to frame K, the photons are continuously reflected and form steady-state electromagnetic waves, which travel to the right (the angle between the laser axis and mirror normal is not exactly zero) and then are released into space. With respect to frame K, zero photons are absorbed by mirrors. SRT predicts that with respect to another Galilean frame, K', which is approaching K with relativistic speed, V_r, the wavelength of photons going toward the upper mirror is red-shifted:

$$\lambda = \lambda_0 \sqrt{\frac{1 + V_r/c}{1 - V_r/c}} > 2\delta_0^{(m)}$$

In addition, with respect to K' the gap between mirrors, δ^m is smaller:

$$\delta^{(m)} = \delta_0^{(m)} \cdot \sqrt{1 - \frac{V_r^2}{c^2}} < \delta_0^{(m)}$$

As a result, $\lambda/2 > \delta^{(m)}$, *i.e.*, with respect to K', the gap between the mirrors is too narrow to allow the existence of such electromagnetic waves. Therefore with respect to K', all n_p photons from each pulse are absorbed by the upper mirror. After absorbing several pulses, the upper mirror overheats and melts. To claim the principle of relativity is valid means to claim $0 = n_p \times m'$ (where m' is the number of pulses required to melt the entire mirror). The upper mirror melts with respect to K' but not with respect to K, which means natural phenomena do not run their course with respect to K' according to exactly the same general laws as with respect to K. By the way, to claim the gap has to exceed half the wavelength with respect to frame K but not with respect to K' is just another way of saying that natural phenomena do not run their course with respect to K' according to exactly the same general laws as with respect to K.

As customary when SRT runs into trouble, it is time to invoke the General Relativity Theory (GRT). Einstein said [4] "All bodies of reference, K, K', *etc.*, are equivalent for the description of natural phenomena (formulation of the general laws of nature)", which means "By application of arbitrary substitutions of the Gauss variables: x_1, x_2, x_3, x_4 the equations (which express the general laws of nature) must pass over into equations of the same form".

Frames K and K' are Galilean therefore also Gaussian. No accelerations are involved in this paradox, therefore frames K and K' remain Gaussian at all times. Consider, for example the number N_1 of mirrors atoms that are in a liquid state after the laser emits m" pulses (m" < m'). With respect to frame K, $N_1 = 0$. With respect to frame K', $N_1 = f(x, y, z, t, ...)$, where f is a very complex function with numerous arguments, including initial atom coordinates, initial temperature of each mirror, the heat transfer coefficient of mirror material(s), the size and geometry of any cooling fins placed on the back side of the mirror, *etc.* Therefore, according to GTR, any arbitrary number is equal to zero.

As usual Lorentz' theory does not lead to any contradiction. According to this theory, the upper mirror does not melt because frame K is at rest.

Many other contradictions invalidate the "principle" of relativity, but cannot be presented here due to lack of space. Fock said [9]: "The general principle of "relativity" is impossible under any physical condition". In conclusion, both SRT and GRT are based on a false assumption. Nevertheless, Einstein has proven an important fact: the existence of a privileged frame of reference and the relativity hypothesis, are mutually exclusive [10 and 11]. Therefore the existence of an absolute frame of reference is certain because the relativity hypothesis is false. A question must be answered now: is it possible to detect this frame using current technology?

The Häfele-Keating experiment [12, 13] involved two atomic clocks flown around the world. One was transported eastward and lost 59 ns, while the other was transported westward and gained 273 ns, compared to the ground clock. This experiment indicates a method of determining the absolute frame of reference. In the absence of accelerations, the rate of on-board clocks depends on absolute velocity, not velocity with respect to Earth. Therefore, the absolute frame can be determined using six space probes equipped with atomic clocks; see Fig. 4-6, which shows the satellite trajectories in Geocentric Inertial Coordinates. The first probe is sent toward +x, the second toward -x, the third toward +y, and so on. When the motion of probes is inertial (probe velocities are kept constant most of the time), the rates of on-board clocks are compared to the rate of a reference clock, which is ground based. The absolute velocity of Earth at any given time can be determined based on the rates of these clocks, corresponding probe velocities and the laws of celestial mechanics.

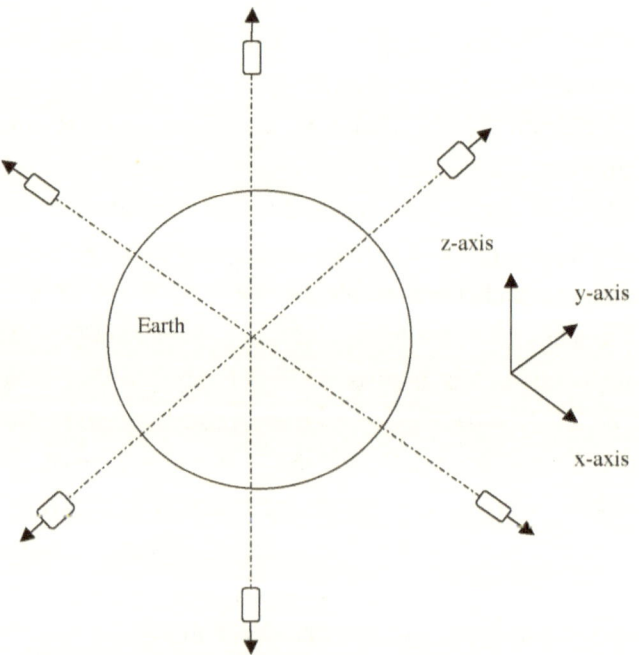

**Figure 4-6. Constellation of space probes suitable for
determining the absolute velocity of Earth**

This experiment could provide very accurate results, but would be very costly and difficult. Fortunately, there are alternatives. For example, Silvertooth used two lasers to measure the standing waves formed by light rays beamed in opposite directions [14–16]. One of the beams was phase modulated with respect to the other in order to create certain phase effects that could be measured using a custom photomultiplier tube. Silvertooth results indicate a consistently privileged direction pointing to the constellation Leo, and a constant velocity of 378 km/sec. In 1992, Silvertooth and Whitney [17] confirmed the results with a subsequent experiment. The absolute frame of reference can also be determined studying the cosmic background radiation.

The goal of the NASA COBE mission was to accurately measure this radiation. It should be noted that COBE was launched after the publication of Silvertooth's results [18]. A Doppler shift revealed a slight anisotropy in the spectrum of the cosmic microwave background. Precise measurements of this anisotropy indicate that the Sun centered frame of reference moves toward the Leo constellation with a velocity of 390 km/sec [18], confirming Silvertooth's results. Therefore the absolute frame of reference has been actually determined and will be exclusively used from now on.

Previous attempts to provide a classical explanation of relativistic phenomena are flawed. SRT and GRT are based on a false hypothesis. The need for a rigorous theory of relativity is now clear. The theory proposed here provides a simple and unitary description of classical, quantum and relativistic phenomena. Being based on quantum dynamics, the relativistic part of this theory is called Quantum Relativity (QR). The name of Quantum Relativity is imposed by tradition. This new theory rejects the principle of relativity, but using another label for relativistic phenomena would create greater confusion.

4.4. The Path toward Physics Reunification

As discussed in Section 1, physics has been broken down into three conflicting branches: Classical Physics, Relativity, and Quantum Mechanics.

Later attempts to reconcile these branches have failed to provide satisfactory results. The conflict between Einstein's relativity and Quantum Theory is insurmountable. To quote: "There is a deep seated fundamental conflict between Quantum Theory and Einstein's Theory of Relativity. Subtle difficulties become insurmountable problems when gravity is added [19]. The root cause of these conflicts is simple: Nature is unitary. Attempts to break down physics into narrow and separated branches, such as Relativity and QM, lead to disjointed distortions of reality. Due to scope, a quantum theory must deal with particles. Einstein's theories of relativity are based on the field concept, *i.e.*, are irreconcilable with any quantum theory by design. In addition, Einstein emphatically dismissed any possible correlation between quantum and relativistic phenomena: "The Theory of Relativity leads to the same law of motion without requiring any special hypothesis whatsoever as to the structure and the behavior of the electron" [4] and thereby indicted both SRT and GTR in very strong terms. The successful reunification of physics requires a theory that is general from inception, not haphazard attempts to patch mutually exclusive aberrations.

The Theory of Internal Energy (TIE) is general from inception, and effortlessly reunites Classical and Quantum Physics. TIE demonstrates that phenomena such as energy quantization, tunneling and radioactive decay are as classical as Newton's laws. TIE proves the particle-wave duality is caused by internal waves, *i.e.* oscillations of particle components within the particle matrix, and is a trademark of complex structures. It is interesting to note that TIE provides a rigorous criterion for the immediate rejection of any future claims regarding the discovery of fundamental particles. There have been way too many such claims in recent history: atoms, "elementary" particles, quarks, strings, etc. It is time to stop this epidemic, and TIE provides a cure, a simple test to discriminate composite objects: all composite objects exhibit wave features; therefore fundamental particles cannot exhibit any wave characteristic. As mentioned, TIE is also applicable to relativistic phenomena.

Note: according to string theory, vibrating, therefore deformable objects are fundamental particles and energy is not conserved. QM and string theory are based on the same mythological non-sense.

4.5. Particle Terminology

The structure of "elementary" particles remains a subject of debate. According to the Standard Model [20] "elementary" particles include "substance-like" components called quarks (or fermions), bound together by force carriers called gauge bosons.

It would be a mistake to link Quantum Relativity to any debatable detail. As discussed in Section 4.2, a link with any debatable idea is a recipe for disaster, even when the theory is essentially correct! In addition, the adopted terminology has to be clear, simple and free of implicit connections to unproven concepts. Therefore the "substance-like" components will be simply called components. Gauge bosons will be called inner quanta (see also Section 4.6). This terminology is simple, based on accepted concepts, free of debatable connotations, and therefore not expected to undermine Quantum Relativity.

4.6. The Hypotheses of Quantum Relativity

As discussed in Section 4.2, the selection of hypotheses is the most dangerous step in the development of any theory and requires the utmost care. To start with, an honest theory should include no tacit hypotheses. It is easier to develop useful new theories when the weak points of current ones are clearly spelled out.

Quantum Relativity is based on three hypotheses:

1) Forces propagate through quanta. Numerous experiments have validated the assumption that electromagnetic forces propagate through quanta and physicists agree that all forces must be related. Furthermore, no one has been able to provide an alternate explanation that is logical.

2)   Components of "elementary" particles are in continuous motion. In order to reject this hypothesis, one must provide examples of perfectly rigid systems and should explain how this rigidity is maintained despite permanent interactions with surrounding objects, which move and therefore exert variable forces on said systems.

3)   The velocity of inner quanta is equal to the velocity of light. All experimental data and theories of force support this hypothesis and justify the adopted terminology. This assumption plays a secondary role in QR. Future experiments that might invalidate this assumption will justify only an adjustment of QR not outright rejection, see Section 4.7 for more details.

Quantum Relativity also relies on the Law of Energy equipartition [21] and three facts:

1)   The existence of an absolute frame of reference (see Section 4.3).

2)   The quantization of energy, a result predicted by both TIE and Quantum Mechanics and experimentally validated.

3)   The periodic motion of components with respect to the particle matrix. If the motion of a component is not periodic, the structure and therefore the integrity of the "elementary" particle will be compromised. Consider for example the mutual distance between components A and B. Assume this distance cannot be represented by a discrete sum of periodic oscillations. This assumption leaves only two alternatives: the distance tends to infinity or to zero (*i.e.*, the components dissociate or merge). In either case; the "elementary" particle is transient, and this version of Quantum Relativity is restricted to stable particles.

"Elementary" particles are assemblies of periodic oscillators, which interact, and according to Fourier analysis, any periodic motion can be represented by a discrete sum of periodic oscillations [21 and 22]. A new idea, *i.e.*, that "substance-like" components are assemblies of quanta (called luxeons), provides an attractive way of simplifying Quantum Relativity; see Section 4.7 for details. But this idea is debatable, and therefore will not be used. Thus, Quantum Relativity is free of any particular hypothesis, unlike all other theories of relativity, (see also Section 4.9).

4.7. The Link between Quantum and Relativistic Phenomena

The force carriers, which assure the cohesion of "elementary" particles are wave quanta; therefore are affected by the Doppler Effect. This effect is the link between quantum and relativistic phenomena. Two Doppler Effects were previously distinguished: classic and relativistic. Only the classic effect is considered here.

The Doppler Effect explains the correlation between the velocity of an "elementary" particle and the energy of all internal oscillators. Inner quanta are emitted by sources moving within an "elementary" particle. With respect to the absolute frame of reference; the frequency of an inner quantum is:

$$f = \frac{c \cdot f_0}{c - \vec{n} \cdot (\vec{V} + d\vec{V})} \qquad (4\text{-}11)$$

where $\vec{V}$ is the particle velocity, $d\vec{V}$ is the source velocity with respect to the particle matrix and is represented by a discrete sum of periodic fluctuations, $\vec{n} = n_1\vec{i} + n_2\vec{j} + n_3\vec{k}$ is a unit vector indicating the quantum path, and f_0 is the frequency of inner quanta emitted by sources at rest. Fluctuation components dV_i are sums of periodic oscillations. The time-averaged value of each component is therefore zero. Furthermore:

$$< \left(dV_i \right)^N > = 0 \qquad (4\text{-}12)$$

Assuming a free "elementary" particle, equation (4-11) can be expanded in a Taylor series as follows:

$$f = \frac{c \cdot f_0}{c - V \cdot \cos(\alpha)} - \frac{c \cdot f_0 \cdot n_1}{[c - V \cdot \cos(\alpha)]^2} dV_1 + \ldots + (-1)^{N-1} \frac{c \cdot f_0 \cdot n_1^{N-1}}{[c - V \cdot \cos(\alpha)]^N} (dV_1)^{N-1} +$$

$$\ldots (-1) \frac{c \cdot f_0 \cdot n_2}{[c - V \cdot \cos(\alpha)]^2} dV_2 + \ldots$$

$$(4\text{-}13)$$

where α is the angle between quantum and particle velocity (see Fig. 4-7).

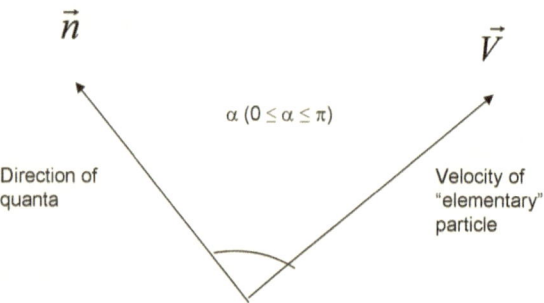

$\vec{n}$ $\vec{V}$

$\alpha \ (0 \leq \alpha \leq \pi)$

Direction of quanta   Velocity of "elementary" particle

Figure 4-7. Angle between the velocity of "elementary" particle and quantum velocity.

In 3D Euclidian space, the angle α formed by two intersecting lines satisfies the condition: $0 \leq \alpha \leq \pi$. Therefore, the average frequency of inner quanta is given by:

$$<f> = \frac{1}{\pi} \frac{1}{t} \int_0^\pi \int_0^t f d\tau d\alpha$$

$$(4\text{-}14)$$

From equation (4-13) and (4-14):

$$<f> = \frac{1}{\pi} \int_0^\pi d\alpha \int_0^t \frac{f_0 d\tau}{1 - V \cdot \cos(\alpha)/c} - \frac{1}{c^2 \cdot \pi} <dV_1> \int_0^\pi \frac{c \cdot f_0 n_1}{[1 - V \cdot \cos(\alpha)/c]^2} d\alpha +$$

$$\ldots + (-1)^{N-1} \frac{1}{c^N \cdot \pi} <(dV_1)^{N-1}> \int_0^\pi \frac{c \cdot f_0 \cdot n_1^{N-1}}{[1 - V \cdot \cos(\alpha)/c]^N} d\alpha + \ldots$$

$$(4\text{-}15)$$

From equations (4-12) and (4-15):

$$< f > = \frac{1}{\pi} \frac{t}{t} \int_0^\pi \frac{f_0}{1 - V \cdot \cos(\alpha)/c} d\alpha = \frac{1}{\pi} \int_0^\pi \frac{f_0 d\alpha}{1 - V \cdot \cos(\alpha)/c} \qquad (4\text{-}16)$$

Using the notation a = 1 and b =-V/c, equation (4-16) becomes:

$$< f > = \frac{f_0}{\pi} \int_0^\pi \frac{d\alpha}{a + b \cdot \cos(\alpha)} \qquad (4\text{-}17)$$

From tables of integrals [23]:

$$\int_0^\pi \frac{d\alpha}{a + b \cdot \cos(\alpha)} = \frac{\pi}{\sqrt{a^2 - b^2}} \qquad (4\text{-}18)$$

From equations (4-17) and (4-18):

$$< f > = \frac{1}{\pi} \frac{\pi \cdot f_0}{\sqrt{a^2 - b^2}} = \frac{f_0}{\sqrt{1 - \frac{V^2}{c^2}}} \qquad (4\text{-}19)$$

Therefore, the average energy of inner quanta, $<E_q>$, is given by:

$$< E_q > = h < f > = \frac{h f_0}{\sqrt{1 - \frac{V^2}{c^2}}}$$

If the luxeon hypothesis is not used, the average energy of components has to be determined otherwise. According to the theorem of energy equipartition, all interacting oscillators have the same average energy [21]. The theorem of energy equipartition was derived for material oscillators and according to TIE, the components of "elementary" particles are indeed material oscillators. Nevertheless, the derivation of the theorem involves only the formal properties of the equation of motion [21]. Any other system, which formally acts like a material oscillator, must have the same equilibrium distribution of energy. Therefore, the law of energy equipartition is also consistent with a description based on Quantum Mechanics. The overall energy of a system is the sum of component energies; therefore the relativistic energy of a free "elementary" particle is given by:

$$E = nh < f > = \frac{nhf_0}{\sqrt{1 - \dfrac{V^2}{c^2}}}$$

$$(4\text{-}20)$$

where n is the total number of components and inner quanta within the particle. With a change of notation, *i.e.*, $m_0 \equiv nhf_0 / c^2$, equation (4-20) takes a well known form:

$$E = \frac{m_0 c^2}{\sqrt{1 - \dfrac{V^2}{c^2}}}$$

$$(4\text{-}21)$$

The relativistic mass is therefore

$$m = \frac{m_0}{\sqrt{1 - \dfrac{V^2}{c^2}}}$$

$$(4\text{-}22)$$

In conclusion, mass is a measure of internal energy. The inner quanta act like springs and components play the role of masses linked by springs. Therefore, mass increases due to the ramping up of internal oscillations.

Particle including substance-like components cannot exist without inner quanta and these quanta do not stop when the particle considered as whole is at absolute rest. Therefore, such particles have rest mass, unlike energy quanta.

Note: Sandu Constatin [24] has shown that equations (4-20) through (4-22) can also be derived based on the Relativistic Doppler Effect.

For accelerating particles, equation (4-13) has to be modified to include acceleration dependent terms and therefore equation (4-20) takes a different form. The physics of such particles will be analyzed in a subsequent paper.

Lorentz has already proved that mass variation, length contraction and time dilation are correlated. Therefore mass increase leads to time dilation (or change in clock rates more exactly) and length contraction.

Quantum Relativity explains the classical Lorentz-Poincare transformation in a rigorously physical manner. All experimental results support Lorentz's theory and therefore Quantum Relativity, which is basically the corrected and updated version of Lorentz' theory. Most experimental results also support SRT and GRT because these theories incorporate Lorentz's transformation. SRT and GRT provide numerous exact and virtually exact predictions with respect to frames at absolute rest and slow-moving frames, respectively. SRT and GRT accurately predict Doppler shifts and the results of length measurements with respect to any frame of reference. But as proven by experiments, Einstein's relativity fails in some instances—remember Dingle's paradox, the Sivlerthooth and Whitney experiment (see Section 4.3) and the Sagnac effect. In agreement with a variety of published results [25–28], QR predicts the genuine speed of light in vacuum is not a constant (because light velocity obeys the classic theorem of velocity addition), but may appear so when the measuring apparatus is affected by relativistic effects.

QR resolves all relativistic paradoxes. Unless at absolute rest, each object suffers some length contraction, mass increase and time dilation. The equation $X - Y = a$ is not sufficient to determine X. A second equation is required. Therefore the relative velocity of two objects is insufficient to determine the Lorentz factors of both. Sometimes the absolute velocity of one object is negligible compared to the absolute velocity of the other. This condition provides the second equation mentioned above. Knowing which object is basically at absolute rest, the relative velocity is sufficient to determine the Lorentz factors of both, but relative velocity alone can be misleading. For example, the absolute velocity of tunnel is negligible; therefore, measurements with respect to Kt provide the true Lorentz factor of the approaching ship-but measurements with respect to Ks are distorted because of intense relativistic effects. As a result, SRT predictions with respect to Ks are wrong. QR clarifies other important issues. Empty space cannot be warped, only matter suffers relativistic changes. Gravity and other fields may also change the energy of inner quanta, therefore the mass of objects. The energy of electric fields surrounding charged particles does indeed increase with velocity due to the Doppler Effect but this increase is not the root cause of relativistic phenomena. QR agrees with all experimental data and unlike all alternatives includes no particular assumptions.

Some mass measurements might depart from equation (4-22) if a significant number of internal quanta are emitted in a preferential direction, for example within an "elementary" particle shaped like a membrane or a string. Three dimensional "elementary" particles might also exhibit such deviations (if quarks are indeed the ultimate building blocks of matter) due to the inherent limitations of statistics based on just a few samples. But hundreds of samples or more allow accurate statistical predictions. Therefore the measured mass of atoms, and molecules, not to mention macroscopic bodies, always satisfies equation (4-22); unless the 3rd QR assumption is far from valid or the ambient gravitational field is strong. As a result Quantum Relativity also offers a criterion for aiding in the search for fundamental particles, which complements the one presented in Section 4.4.: if the measured mass of a structure always satisfies equation (4-22), the structure is at least two levels above the ultimate building blocks of matter.

Einstein himself considered that theories based on fields might be superseded by particle theories [29]: "I consider it quite possible that physics cannot be based on the field concept, *i.e.* on continuous structures. In that case, nothing remains of my entire castle in the air, gravitation theory included, and of the rest of modern physics". Newtonian Mechanics is based on particles. Therefore, Classical Physics is partially based on particles. Quantum Relativity and TIE are also based on particles not fields and merge seamlessly with Classical Physics. These three theories form a consistent and broad foundation for a General Theory of Physics.

4.8. Unresolved Issues

Marmet [30] analyzed in more detail the correlation between relativistic energy, length contraction, and clock rates. He also analyzed the correlations between etalons of mass and length placed in different frames of reference, proposed a coherent terminology for various relativistic parameters and demonstrated that the relativity "principle" violates the Law of Energy Conservation. Poincaré showed that a real length contraction requires a real change in clock rates [31, 32]. Marmet also explained the advance of Mercury perihelion without using the relativity "principle" [30].

Nevertheless, many issues remain unresolved. Here are just a few. The impact of mass increase on the structure of atoms and molecules is not entirely clear. At first glance, the coefficient of lateral contraction appears to be exactly 1 because transversal motion has no impact on the frequency of inner quanta. But macroscopic bodies contain a multitude of "elementary" particles involved in non-inertial motion and complex interactions. Beckmann [33] thinks electromagnetic waves and gravity are very closely related. This interesting idea should be carefully investigated and leads to two important questions: are there other factors, which might affect the speed of light and inner quanta and therefore relativistic energy? Is it possible to manipulate the speed of inner quanta using a combination of such factors? Such control could allow the development of materials with enormous strength, new medical treatments, quantum leaps in computer performance, faster than light starships and who knows what else.

4.9. In Defense of Quantum Relativity

Because the ultimate fate of a theory is dictated by the quality of underlying hypotheses (see Section 4.2), here is a review of particular assumptions included in various relativity theories. A particular hypothesis is a critical weakness in any theory. A theory that is free of particular hypotheses is far more likely to endure because it is much harder to reject a group of theories than an isolated one. For example Quantum Relativity can be rejected only together with current theories of quantum phenomena. Competing theories of relativity rely on particular hypotheses and therefore stand alone. A theory based on one particular hypothesis picked from two alternatives has a chance of failure of 50%. A theory based on two particular hypotheses has a chance of failure of 75% and so on.

In general, aether theories rely on three or more particular hypotheses. To assume the existence of aether is just the first hypothesis. Various other assumptions must be included. For example, aether is all-pervasive, entrained by bodies, exerts no drag while motion remains inertial and the speed of light is constant with respect to aether [33]. The assumption that aether is some kind of fluid leads to a contradiction. Drag exerted on inertial bodies can be zero only if circulation is zero, which means the aether "fluid" has null viscosity [34]. The hypotheses of null viscosity and of aether entrainment are mutually exclusive.

Table 4-1 provides a list of particular hypothesis included in theories of relativity discussed here and the associated risk of failure. Being the only theory free of particular hypotheses, Quantum Relativity is the most likely to endure.

4.10. Conclusions and Recommendations

Relativistic energy and velocity are correlated due to the Doppler Effect. This correlation causes length contraction and time dilation or more exactly a

Table 4-1. Particular hypotheses included in competing
theories of relativity and the associated risks of failure.

Theory	Hypotheses	Risk of failure due to hypotheses	Comments
Special Relativity Theory	1. "If relative to K, K' is a uniformly moving co-ordinate system devoid of rotation, then natural phenomena run their course with respect to K' according to exactly the same general laws as with respect to K. This statement is called the principle of relativity (in the restricted sense)" [4] 2. The velocity of light in vacuum is a constant in all inertial frames of reference 3. The units of measure do not change as a result of transfer from one frame of reference to another. 4. The transversal contraction factor is exactly equal to 1.	93.75%	The first hypothesis is false. The number of experiments contradicting SRT is increasing.
General Theory of Relativity	1. "By application of arbitrary substitutions of the Gauss variables x1, x2, x3, x4, the equations (which express the general laws of nature) must pass over into equations of the same form" [4]. 2. The units of measure do not change as a result of transfer from one frame of reference to another.	75%	The first hypothesis is false.
Aether theories	1. There is an undetected medium called aether 2. Aether permeates all space 3. Aether does not exert any force on inertial objects…	87.5% or more	There is no experimental data to support the aether hypothesis. Classic Physics explains relativistic phenomena without this hypothesis.
Quantum Relativity	None	0%	Agrees with all experimental data

change in clock rates. The speed of light in vacuum is not constant with respect to all inertial frames but may appear so whenever the experimental apparatus is not designed to compensate for length contraction and time dilation.

Unlike competing theories, Quantum Relativity, which is organically related to Classical Physics, explains relativistic phenomena without recourse to any particular hypothesis. Classical Physics, Quantum Relativity and the Theory of Internal Energy form a rational, consistent and simple paradigm for classical, quantum and relativistic phenomena, and a broad basis for a General Theory of Physics.

References

1. A.A. Michelson and E.W. Morley, "On the Relative Motion of the Earth and the Luminiferous Ether", Amer. J. Sci. 34, 1887, pp. 333–345.

2. A.A. Michelson and E.W. Morley, "On the Relative Motion of the Earth and the Luminiferous Aether." Philos. Mag. 24, 1887, pp. 449–463.

3. H.A. Lorentz, "Electromagnetic phenomena in a system moving with any velocity less than that of light", Proceedings Academy of Sciences, vol IV, 1904, pp. 669–678.

4. A. Einstein, Relativity, "The Special and General Theory", Methuem and Co Ltd, 1920, pp. 15, 23, 24, 35, 36, 42, 48, 50, 55, 83, 111.

5. Born, M, "Zur kinematics des starren Körper im System des Relativitätsprinzips", Göttinguer Nach., 1910, pp. 161–179.

6. P. Marmet, "Fundamental Nature of Relativistic Mass and Magnetic Fields", International IFNA-ANS Journal "Problems of Nonlinear Analysis in Engineering Systems", Vol. 9, 2003, No. 3 p. 19.

7. D. Bohm, "A Suggested Interpretation of the Quantum Theory in Terms of Hidden Variables, I and II", Physical Review 85, 1952, pp. 166–193.

8. H. Dingle, "Science at the Crossroads", Martin Brian and O'Keefe, London, 1972.

9. V. Fock, "Theory of Space, Time and Gravitation", Pergamon Press, London, 1959, p.401.

10. J. Lévy, "Is the relativity principle an unquestionable concept of physics?", Physical Interpretations of Relativity Theory, (P.I.R.T), Imperial College London, late papers, 1998, pp. 156, 401

11. Lévy, J., "Critique of some assumptions of special relativity and arguments in favor of an aether frame", Physical Interpretations of Relativity Theory VII, Imperial College London, September 2000, pp. 15–18.

12 J.C. Häfele, and R.E. Keating, "Around-the-world Atomic Clock: Predicted Relativistic Time Gains", Science, 177, 1972, pp. 166–167.

13. J.C. Häfele, and R.E. Keating, "Around-the-world Atomic Clock: Measured Relativistic Time Gains", Science 177, 1972, pp. 168–170.

14. E.W. Silvertooth, "Special Relativity", Nature 322, August 1986, p. 590.

15. E.W. Silvertooth, "Experimental Detection of the Ether", Speculations in Sci. and Techn. Vol.10, 1987, p. 1.

16. E.W. Silvertooth, "Motion Through the Ether", Electronics & Wireless World, May, 1989, pp. 437–438.

17. E.W. Silvertooth, and C.K. Whitney, "A New Michelson-Morley Experiment", Physics Essays, 5, 1992, pp. 1, 82–88.

18. W.H. Cantrell, (editorial): "A Dissident View of Relativity Theory", Infinite Energy 10. (59), 2005, pp. 6–13.

19. P. Renteln, "Quantum Gravity", American Scientist 79, 1991, pp. 508–527.

20. David J. Griffiths, "Introduction to Elementary Particles", John Wiley, & Sons, Inc, 1987, pp. 37–51.

21. D. Bohm, "Quantum Theory", Dover Books, Dover Publications, Inc., N.Y., 1979, pp. 10, 16–17.

22. C.M. Harris "Shock and Vibration Handbook", 4th edition, McGraw-Hill, 1961, pp. 22.10 and 22.11.

23. S.M. Selby (editor in chief), CRC, "Standard Mathematical Tables", 19th edition, The Chemical Rubber, Co., 1971, p. 447.

24. Sandu C. and Brasoveanu D., "Sonic-Electromagnetic-Gravitational-Spacecraft, Part 4—Doppler Effects, Inertia and Mass Increase", Proceedings of the AIAA SPACE 2007 Conference & Exposition, Long Beach, California, September 18–20, 2007.

25. C. Renshaw, "Explanation of the Anomalous Doppler Observations in Pioneer 10 and 11," Proc. IEEE Aerospace Conf. 2, 1999, pp. 59–63.

26. B.G. Wallace, "Radar Testing of the Relative Velocity of Light in Space", Spectroscopic Letters **2**, 1969, p. 361.

27. B.G. Wallace, "Letter to the Editor", Physics Today 36, 1983, p. 1.

28. B.G. Wallace, "The Farce of Physics", 1994, online at http://surf.de.uu.net/bookland/sci/farce/farce_toc.html.

29. A. Pais, "Subtle is the Lord, The Science and the Life of Albert Einstein", Oxford University Press, Oxford, UK, 1982, p. 467.

30. P. Marmet, "Einstein's Theory of Relativity versus Classical Mechanics", Newton Physics Books, 1997, pp. 27, 28, 81–92.

31. H. Poincare, "Les limites de la loi de Newton", Bulletin astronomique, Observatoire de Paris, 1906–1907, Gauthier-Villar, XVII, p. 1953.

32. H. Poincare, "La dynamique de l'électron", Revue Générale des Sciences Pures et Appl. 19, 1908, pp. 386–402.

33. P. Beckmann, "Einstein Plus Two", The Golem Press, 1987, p. 27–28.

34. J. Florea, and V. Panaitescu, Mecanica Fluidelor, Editura Didactica si Pedagogica Bucuresti, 1979, pp. 164–166, 200–201.

978-0-595-48499-7
0-595-48499-9

www.ingramcontent.com/pod-product-compliance
Lightning Source LLC
Chambersburg PA
CBHW030413290526
45785CB00004B/1983